# INSTRUCTOR'S MANUAL

TO ACCOMPANY

# ELECTRIC AND MAGNETIC INTERACTIONS

RUTH W. CHABAY

BRUCE A. SHERWOOD

Carnegie Mellon University

JOHN WILEY & SONS, INC.

New York Chichester Brisbane Toronto Singapore

ISBN 0-471-11467-7

Printed in the United States of America

10 9 8 7 6 5 4 3 2 1

Printed and bound by Malloy Lithographing, Inc.

# CONTENTS


## 1: A new approach to E&M 1

Purpose of this Instructor's Manual 2
Problems with the traditional E&M course 2
Goals and approach of the new course 3
Components of the course 4
Student expectations and attitudes 8
Evaluation 9
Detailed commentary on individual chapters 10
How we run our own course 17
Grading qualitative questions 19
References on surface charge in circuits 21

## 2: Procurement information for desktop experiments 23

## 3: Course description for students 27

## 4: Sample assignment sheets for students (semester A) 31

## 5: Daily log of class activities (semester A) 47

Structure of a typical week 48

## 6: Sample quizzes & exams (semester A) 67

## 7: Sample quizzes & exams (semester B) 91

## 8: Formal labs 117

Circuit lab with multimeters 118
Magnetic induction lab 129

## 9: Solutions to selected homework problems 141

# CHAPTER 1

# A NEW APPROACH TO E&M


| | |
|---|---|
| Purpose of this Instructor's Manual | 2 |
| Problems with the traditional E&M course | 2 |
| Goals and approach of the new course | 3 |
| Components of the course | 4 |
| Student expectations and attitudes | 8 |
| Evaluation | 9 |
| Detailed commentary on individual chapters | 10 |
| How we run our own course | 17 |
| Grading qualitative questions | 19 |
| References on surface charge in circuits | 21 |

## Purpose of this Instructor's Manual

The electricity and magnetism course for which our book is designed differs significantly from the traditional introductory E&M course in sequence, content, and pedagogical approach. We ourselves have now taught this course for seven semesters, using increasingly refined draft versions of the book. The published version of the book reflects much of what we have learned about how to teach this material effectively.

However, not all of what we know about teaching this course can be incorporated into materials intended for student use. We have prepared this Instructor's Manual in order to present explicitly some of the pedagogical insights we have gained in developing and teaching this course, which are likely to be useful to other teachers. Since the logistics of running a course are at least as important as the pedagogical details, we also provide in this manual the following:

- procurement information for the desktop experiments;
- a course description given to our students;
- assignment sheets given to our students;
- a detailed log of daily class activities in lecture and recitation;
- quizzes, exams, and final exams from two different semesters;
- formal laboratory experiments on circuits and magnetic induction; and
- solutions to selected homework problems.

## Problems with the traditional E&M course

Both teachers and students have long suspected that many students emerge from well-taught introductory physics courses with little *physical* understanding. Systematic research over the past decade has confirmed these suspicions; students who are able to solve traditional "quantitative" problems often cannot answer straightforward qualitative questions based on the most fundamental aspects of the science. For example, many students after doing well in a good traditional course on electricity still believe that electrons are used up in a flashlight, being converted into light in the bulb! Physical understanding is not an automatic side effect of doing lots of quantitative problems, especially if this problem solving is based on the manipulation of secondary formulas rather than being based on reasoning from fundamental principles.

This problem is common to all topics in the introductory physics sequence. However, several characteristics of the traditional introductory E&M course pose special difficulties.

First, the traditional approach to E&M is often highly formal and mathematical. This emphasis on formal mathematics reinforces students' belief that the course is essentially an applied math course, and that mathematical manipulation is the central goal. Qualitative reasoning is not taught explicitly.

Second, many layers of abstract concepts are introduced in a very short time, too quickly for most students to digest them and distinguish between them. Often charge, electric force, electric field, Gauss's law, and electric potential are introduced within the first three weeks of the course. This is a very heavy conceptual overload, and it is not entirely surprising that students can emerge from the course with unresolved confusions between field and potential, or even between charge and field.

Third, the conceptual framework of the course is fragmented and disjoint. Electrostatics and circuits seem to be separate sciences and require completely different kinds of analyses, based on different sets of concepts (charge, force, and field for electrostatics; potential and current for circuits).

Fourth, many students today are unfamiliar with the basic phenomena studied in the E&M course. Typically they have never observed electrostatic repulsion, nor assembled a simple circuit, nor deflected a compass with a magnet, nor made an electromagnet by winding wire around a nail.

Consequently, it is not clear to them *why* we are constructing an elaborate theoretical framework, or what it is supposed to explain.

Finally, there is typically very little twentieth century physics in the course. This is true of other components of the introductory physics sequence, but it is particularly striking in E&M courses, which sometimes go to great lengths to avoid mentioning matter, and to avoid discussion of relativistic effects. This is a missed opportunity, since students are already somewhat familiar with atoms, molecules, electrons, and protons, which can provide a concreteness that helps offset the excessive formality of the subject, and they are intrigued by relativity and retardation effects.

## Goals and approach of the new course

The course embodied in this book addresses the problems outlined in the previous section. Our goals in the development of a new, calculus-based introductory E&M course have been:

- to engage the students in a process central to science—the attempt to explain in detail a broad range of electric and magnetic phenomena using a small set of powerful fundamental principles;
- to base instruction on "minimalist" pedagogical principles;
- to teach explicitly qualitative as well as quantitative reasoning;
- to stress an atomic-level analysis of the interaction of fields and matter; and
- to involve students actively in observing and analyzing actual phenomena.

### *A small set of fundamental principles*

Using a few fundamental principles to explain a broad range of phenomena is characteristic of much of the day-to-day work in science. Scientists are often portrayed as being constantly engaged in discovering new physical laws, but such activities are in fact quite rare, and it is very difficult to engage students honestly in discovering (known) physical laws for themselves. What we can do is engage students in observing and explaining phenomena in the minimalist and reductionist style common to much of real physical science as it is actually practiced.

### *"Minimalist" pedagogy*

This minimalist view of science is echoed by our minimalist approach to pedagogy: We normally introduce a topic only if there is an associated task such as an experimental measurement or a homework problem that the student can carry out, and we omit topics if there is no appropriate related task for the student. This philosophy leads to removing much of the clutter that can prevent students from seeing the central aspects and the unity of the subject matter. What remains after getting rid of less central topics is likely to be more coherent and memorable, with a heightened probability that a year later students might still know a lot about electric and magnetic interactions.

#### Unifying electrostatics and circuits

A striking and distinctive consequence of the emphasis on a small number of fundamental principles and a minimalist approach to pedagogy is that we unify the treatment of electrostatics and circuits by initially introducing just enough principles (charge, electric field, properties of metals) *to permit analyzing circuits in the same way that we analyze electrostatic phenomena*, directly in terms of the fundamental Coulomb interaction. We delay introducing the more abstract electric potential until later, after which we then re-analyze circuits in terms of potential. This radical approach to circuits has far-reaching consequences on the overall structure of the course, including the sequence of topics.

### *Qualitative reasoning*

We give as much emphasis to qualitative reasoning as to the usual quantitative reasoning. Qualitative reasoning is important in its own right, because it is often the path to deep physical understanding. Moreover, qualitative understanding provides a conceptual framework within which to carry out quantitative analyses: The equations make more sense.

However, qualitative questions given in isolation can appear to students as merely tacked on and not representing "real" physics. We integrate qualitative and quantitative analyses throughout the book, and we require the students to pay the same serious attention to both kinds of work Performance on exams suggests that the emphasis on qualitative reasoning has led to improved work on quantitative problems, with a lower rate of major conceptual errors and errors in the choice of method of attack.

### *Emphasis on matter at the atomic level*

In addition to a strong emphasis on qualitative reasoning to complement and support quantitative reasoning, the new course places heavy emphasis on the atomic structure of matter. For example, while electric polarization is often a minor or optional topic in introductory texts, for us it is a vital, central topic. Early on we make major investments in understanding the differing mechanisms for polarization in insulators, metals, and ionic solutions (especially, the salt-water layer on the skin). This emphasis on atomic-level description and analysis is useful in its own right, because it continues the student's long developmental progress toward a mature conception of the nature of matter, a process already begun in introductory chemistry courses. It also brings a contemporary world view into a course which too often has felt like an exercise in 19th-century potential theory.

Reasoning concretely in terms of atoms and subatomic particles is very often much easier and much more vivid than attempting to reason solely in terms of macroscopic continuum concepts such as "charge" or "current." For example, charge conservation is much easier to think about in terms of counting elementary charged particles than in terms of unspecified electric "stuff." This advantage of microscopic models is similar to the advantage of using the kinetic theory of gases to illuminate the macroscopic and formal aspects of thermodynamics.

### *Observing and analyzing real phenomena*

Throughout the course we integrate theory and experiment by the use of simple but profound "desktop" experiments which can be done in the regular classroom or at home.

## Components of the course

The new course has the following major components which support our major goals:

- an interactive student textbook, *Electric and Magnetic Interactions*;
- desktop experiments that can be carried out at home or in the regular classroom, and which are tightly integrated with the theoretical treatment (whereas physics labs are often quite divorced from the lectures and text);
- a restructured curriculum, delaying some particularly difficult topics such as potential and Gauss's law until the students have the necessary fundamental understanding of charges and fields, and unifying the treatment of electrostatics and circuits;
- homework problems that require complete written scientific explanations; and
- educational software to add physical insights unavailable by other means.

While the book is written for the calculus-based college course, some high school physics teachers have used the book successfully by judiciously skipping certain sections.

## An interactive textbook

It is often the case in introductory physics courses that students do not really study the body of a textbook chapter but simply flip through it when doing the homework problems, looking for useful secondary formulas. This contributes to a lack of fundamental physical understanding and an emphasis on formula matching. Other students read the text passively, without pausing to try to apply concepts or check their own understanding. To address this problem, *Electric and Magnetic Interactions* is designed to encourage students to read the material, become engaged with it, and work actively on it.

The textbook is interactive: there are empty boxes where the student is asked to give an explanation, complete a derivation, or record an experimental measurement. There is an emphasis on qualitative reasoning that leads into and supports the quantitative work. Solutions at the end of each chapter are written as though a student had written them, to provide models of what good student explanations should look like. In some cases the solutions also present incorrect or inadequate explanations, with appropriate commentary aimed at common misconceptions. Each chapter has the following structure:

- table of contents and overview of the chapter
- main body, with theory and experiments intimately intertwined
- brief summary of the main points (typically one page)
- review questions, which induce the student to reflect on the main points of the chapter
- homework problems, designed to support the approach by emphasizing the integration of diverse concepts and by de-emphasizing reliance on formula matching
- model student solutions for the activities in the main body of the chapter

## Desktop experiments

An important feature of the book is the use of integrated desktop experiments, intertwined with the theoretical treatment, which the students can do in a regular classroom or at home. Each student *must* have his or her own kit of simple equipment, to perform desktop experiments that delve very deeply into the fundamental atomic nature of electric and magnetic interactions. These experiments deal with static electricity, electric circuits, and magnetism. The desktop experiments can be supplemented by formal lab sessions in which students carry out experiments that are beyond the scope of the desktop equipment (examples of two such formal labs are included in this manual). The logistics of kit ownership can be worked out in a variety of ways; students may buy them, or borrow them from the department. Unless a great deal of classroom time is set aside for working through the book, it is important that students be able to take the equipment home, so that they can do experiments at any time convenient to them, while simultaneously working on the theory.

Preassembled kits are available from the publisher. If you prefer to assemble your own, see the detailed procurement information included later in this manual. The kit includes a roll of "invisible" tape (generic invisible tape seems to be better for electrostatics experiments than Scotch™ brand Magic™ Tape), two D cells, a battery holder, two flashlight-bulb sockets, two high-resistance (#48) and two low-resistance (#14) flashlight bulbs, seven clip leads, a 2-meter length of hook-up wire, 50-cm lengths of #26 and #30 nichrome wire, a liquid-filled magnetic compass, a half-farad capacitor (needed for Chapter 7), a small bar magnet, and three initially unmagnetized iron nails.

## A restructured curriculum

A typical sequence of topics for the introductory E&M course consists of charge, electric field and Gauss's law, potential, circuits, magnetic force, magnetic field and Ampere's law, magnetic induction, and electromagnetic radiation. We have re-ordered the topics in a way that makes more pedagogical sense.

### Field as the unifying theme

The unifying thread running through all the topics is the concept of field. As one physics teacher said after using our book, "Usually when we get to circuits we talk only about potential and stop using electric field entirely, so the students forget the field concept. In the new curriculum the students keep getting practice in the use of electric field, which runs through the entire course."

The most radical change we have made is to analyze DC and RC circuits initially in terms of electric field and surface charge on the wires, *before* introducing the more abstract concept of electric potential. This change has far-reaching benefits. It makes circuit behavior far less mysterious and abstract, and it unifies electrostatics and circuits, which usually seem like two separate disciplines. The role of surface charge in contributing to the electric field in a circuit is not widely known in the physics community, despite many articles on the topic in physics journals over the last 50 years (see references later in this manual). To the best of our knowledge, students have never before been carefully and systematically taught how to analyze circuit behavior using the same fundamental principles that are used to analyze electrostatic phenomena.

In addition to unifying electrostatics and circuits, analyzing circuits initially in terms of field rather than potential places additional emphasis on and gives more practice with the concept of electric field, and on the relationship between charge and field. We delay the study of electric potential until the student has a solid understanding of charge and field, which makes it easier to understand potential. After introducing potential, circuits are re-analyzed in terms of potential, but without losing sight of the role of electric field, and the relationship between field and potential. This opportunity to deepen student understanding of field and potential is missed in most E&M courses, where electric field plays little role in connection with circuits.

### Delaying Gauss's law

Another major change in the sequence of topics is to delay the study of Gauss's law until the very end of the electricity portion of the course, just before beginning the study of magnetism. Our rationale for this unusual delay is that without an exceptionally strong grasp of the relationship between charge and electric field, it is nearly impossible to understand Gauss's law. The beginning of the course is far too early to bring up Gauss's law. Moreover, when we finally do discuss Gauss's law, we introduce the topic in an intuitive and visual way that builds on students' knowledge of patterns of electric field in space.

How do we manage to do without Gauss's law throughout the chapters on electricity? First, at all times we emphasize the electric field of finite rather than infinite charge distributions. In particular, rather than emphasizing the field of an infinite plate (with the accompanying pathological infinities, which we avoid), we treat the on-axis field of a uniformly-charged finite disk in great detail, which then gives us not only the field near the center of a parallel-plate capacitor made of two disks but also the fringe field just outside the center of the plates, which turns out to be extremely useful in qualitative reasoning about RC circuits in terms of field rather than potential. As one student commented, "You could never charge or discharge an infinite-plate capacitor, because it would take an infinite amount of time!"

Second, early in the course we state as unproved assertions that any excess charge in a metal is found on a surface (not in the interior), and that a uniformly-charged spherical shell makes a field outside like a point charge and a zero field inside. We show that these assertions are plausible, but of course it requires Gauss's law to prove these properties efficiently, and we do eventually carry out the proofs when we introduce Gauss's law. We believe that the necessity of asking the students for trust on these two points is worth it in terms of the major decrease in complexity that comes from avoiding Gauss's law early in the course.

### Conventional current

A minor but significant sequence issue is that we introduce current in terms of electrons per second in a metal wire rather than in terms of the conventional current of positive carriers measured in coulombs per second. The emphasis on electron flow connects to the earlier work on electrostatics and the polarization of metals, while the description of current in terms of countable electrons per second provides a significantly more concrete and vivid mental image of current than is provided by continuum flow. Of course we do eventually introduce the concept of conventional current, but this comes quite late in the course.

#### Relativity

Special relativity plays a significant role throughout the course. It first comes up in Chapter 3 when we introduce electric field: Retardation effects imply that electric field is not just a mathematical convenience but has an existence of its own, even after the source charges have moved to a new location. It is not necessary that the students have previously studied relativity in a formal way. In our experience, all students have heard that nothing can travel faster than light, and this notion is sufficient for our purposes. Retardation is again invoked in Chapter 6 on electric circuits, to give an order-of-magnitude estimate for how long it takes after making a final connection for the steady state of a DC circuit to be established.

In Chapter 11 on magnetic field we point out that retardation effects imply the existence of magnetic field even after the sources have been moved, just as in the case of electric field. We also point out that there is something odd about the presence of "v" in the Biot-Savart law, because this implies some sort of relationship between electric and magnetic fields, depending on the reference frame.

In Chapter 12 on magnetic forces we discuss in detail the motion of two protons which repel each other, as viewed from a reference frame where the protons are initially at rest (where the forces are purely electric), and from a moving reference frame (where the forces are both electric and magnetic). A straightforward calculation leads to the astounding result that clocks must run at different speeds in the two reference frames. Many students are delighted and intrigued by this discussion.

Retardation again plays a role in Chapter 14 where it underlies a graphical plausibility argument that an accelerated charge will produce a moving transverse electric field.

#### A note on the omission of field lines

It may seem an odd omission that we don't discuss field lines. We searched through many textbooks looking for homework problems whose solution depended on using the concept of field lines, but the only problems we found were problems *about* field lines (can they cross, what does their density signify, do charged particles move along field lines, etc.). We have taken a minimalist approach: If there isn't an important student task that requires a particular concept, then that concept need not be introduced at all. On this basis we have omitted trying to teach about field lines, because although field-line plots can be very pretty, at the introductory level they seem to be hardly used for anything and yet present many opportunities for fostering misconceptions.

### *Complete written scientific explanations*

The book supports our attempts to get students to write clear and complete explanations of phenomena. In previous physics courses the students may have gotten away with merely listing a few equations, but we require the students to justify the equations, including stating any assumptions that must be made to solve a problem, and this requirement is part of quizzes and exams. Initially the students need quite a bit of help in learning what constitutes a good written explanation (including clear and appropriate diagrams), and some explicit assistance is given at the end of Chapter 2.

An understandable initial reaction on the part of the students is to think that we want lots of prose (since they didn't write any in previous courses). It takes persistent effort to show that terse but focused prose is better than lots of prose, and that a good diagram really is worth a thousand words. We sometimes show the students good and poor examples of explanations given by other students (for example, see exercises 2.7.5a and 2.7.5b on page 59 in the book, and the corresponding solutions on pages 81-82). Also note the completeness of the detailed solutions at the end of each chapter (and solutions to selected homework problems given later in this manual, some of which we make available to students), which serve as models for the students of what we consider to be good explanations. Moreover, the homework problems are aimed at engaging the students in physical understanding and complete explanations.

### Computer programs

We have written several computer programs (in the cT programming language) which foster a qualitative feel and provide scientific visualization for abstract concepts. We have used them as lecture demonstration devices and sometimes for individual student assignments. Among the programs are *EM Field* by David Trowbridge and Bruce Sherwood, which won a first prize in the 1991 educational software contest of the journal *Computers in Physics*, and *Electric Field Hockey* by Ruth Chabay, which won a first prize in the 1992 contest. Brief descriptions may be found in an article by Bruce Sherwood and David Andersen in the journal *Computers in Physics* (March/April 1993, pp. 136-143). *EM Field* and *Electric Field Hockey* are available from Physics Academic Software: TASL, P.O. Box 8202, North Carolina State University, Raleigh NC 27695-8202; phone 800-955-TASL or 919-515-7447; fax 919-515-2682.

## Student expectations and attitudes

We have grown increasingly sensitive to critical issues concerning student attitudes toward science and toward science courses. Our students, who are majoring in science or engineering, come to us with serious misconceptions about the nature of science and with attitudes instilled in prior science courses that their task should mainly consist of finding a formula into which to plug some numbers to get an exact numerical answer. They initially feel betrayed that what they have learned about how to approach a science course seems to be somewhat irrelevant, and that we are asking them to think deeply, to reason qualitatively, to make approximations, and to construct rigorous scientific explanations, none of which seems to them to have much to do with what they thought science was about. We invest a lot of time and energy acculturating the students to the real scientific enterprise, and our book is crafted to support these efforts. As a result of explicitly addressing these issues, after a few weeks most students buy into our approach.

This problem of student expectations and perceptions is not unique to our course. There have been reports of students at other universities in innovative science or math courses complaining bitterly, "I'm paying these teachers thousands of dollars in tuition, and they won't tell me the answers to things! Why should I have to work this stuff out when they could just tell me? What am I paying them for, anyway?"

When we first taught this new course, we did not recognize how serious this problem can be, and it is likely that some students went through the entire course without understanding what we were trying to do. Now that we understand the problem, we are very explicit right from the beginning and continually throughout the course in explaining what we are trying to do and why, and this has essentially solved the problem. Most students now respond very well to being asked to think for themselves.

We had a student who during the first few weeks of the course appeared to be exceptionally weak and definitely headed for failing the course. Just before the first exam, she came to office hours to ask for clarification of some points. It was puzzling to find that she *did* seem to understand the main concepts. And then she said, "I'm going to be honest and tell you that for the first three weeks I was really annoyed with this course, and I was really annoyed with you. In previous science courses the teacher told me what I needed to know, but in this course you wouldn't tell me anything. And then I came to realize that you wanted us to figure things out for ourselves! I can see that in fact this is a much better scheme, and now I really get it—I understand what you are trying to do, and I understand the material." She got one of the highest scores on the exam.

However, there are always a few students who never do understand what we are trying to achieve. In an anonymous course evaluation at the end of the semester, one student said, "I couldn't devote myself to the hours necessary to go through and understand this stuff, and I sure couldn't study, because it takes almost 20 minutes to find a formula." It is clear that this student objects to being asked to *understand* the material and believes that "study" should consist of efficient searches for secondary magic formulas that will solve related homework problems. (Note that all of the *fundamental* formulas are given on the summary page at the end of each chapter, and that on the same evaluation form this student estimated spending a mere 4 hours per week studying outside class, where the average was about 7.5 hours.)

*We urge you to pay close attention to these issues with your own students.* A careful explanation of your goals on the first day of the course is necessary but not sufficient: You need to return to these issues repeatedly throughout the course. Our book provides strong support for bringing more science into an introductory science course, but you need to provide additional inspiration and guidance to help students overcome inappropriate attitudes that may have been inadvertently fostered in previous courses.

## Evaluation

How should a new course be evaluated? Unlike the situation with classical mechanics, there are no comprehensive standard evaluation instruments to test physical understanding of E&M. Furthermore, the difference in content and emphasis in this course make comparisons between this and other courses difficult. We have looked at three different measures of success in attempts to evaluate the effectiveness of this course: students' performance on our own exam questions, students' performance on a few qualitative questions for which there are baseline data provided by physics education research, and students' evaluations of their own learning in the course.

### Evidence from exams

We have begun to look in detail at students' work on quantitative problems of a traditional kind, given on final exams. Since we have not had a formal control group, our preliminary analysis is based on similar problems given on final exams in our own course over several years, as more and more of our own materials were introduced into the course. Compared to semesters in which we used a standard textbook and a mostly traditional approach, the work of students in the new course appears to be cleaner and to show a rate of conceptual errors that is considerably lower than is usually the case. There are many fewer cases of invoking irrelevant or incorrect relations, or of garbled reasoning.

### Comparison to baseline data

On a final exam question devised by Lillian McDermott of the University of Washington, students were asked to compare the ammeter readings in three circuits with identical batteries: an ammeter followed by a bulb, a bulb followed by an ammeter, and an ammeter alone. 90% of our students answered the question correctly, whereas McDermott (private communication) reports that only 60% of the students in a standard course with traditional lab answered the question correctly. We believe that our excellent results stem from a heavy emphasis on a microscopic analysis of circuits, including the fact that current conservation in the steady state is the result of charge conservation and dynamic feedback.

On another final exam question devised by McDermott, students were asked to compare the brightness of identical bulbs with identical batteries in three different circuits: one bulb, two bulbs in series, and two bulbs in parallel. In our emphasis on fundamental interactions we place less stress than usual on some practical aspects of resistive circuits, because we feel that much of circuit analysis belong more to technology than to physics, or to electronics courses rather than to this course. Nevertheless, 66% of our students correctly described the various bulb brightnesses, whereas McDermott reports that only 15% of the students in a standard calculus-based course with traditional lab answered the question correctly (see L. C. McDermott & P. S. Shaffer, "Research as a guide for curriculum development: An example from introductory electricity. Part I: Investigation of student understanding." American Journal of Physics, **60**, 994-1003).

### Students' opinions

Most students are enthusiastic about the course. Particularly noteworthy are the many comments that our book is easier to learn from than standard textbooks, and comments that praise the emphasis on fundamental principles. It is particularly gratifying to get questions from students about what courses they could take on relativity or quantum mechanics, because we have deliberately tried to show the students where the classical picture is inadequate. A small number of students do not like the workbook format because they feel that it takes too much time and requires that the student work at understanding the material rather than simply and efficiently being told the facts. We of course see these objections as missing the point of what a science course should be about.

## Detailed commentary on individual chapters

In this section we comment on the ideas we have chosen to emphasize in each section of the course, and on opportunities and student difficulties of which we have become aware in our own teaching of the course.

### *Charges, fields, and superposition*

The most important theme of the book is that charges make fields, and fields affect *other* charges. This conceptual framework is so central to E&M that we emphasize it even when analyzing circuits, which is a major curriculum innovation. A closely related issue is that we repeatedly emphasize the superposition principle for fields, in many different settings. We have found that the superposition principle is not at all obvious to students. For example, students have the intuitively appealing notion that matter blocks fields, and they do not immediately see that this is inconsistent with the superposition principle.

As with many other fundamental issues such as current conservation, it is not sufficient merely to tell students once or twice that matter does not block fields. Rather one must engage the students in the analysis of many different kinds of applications of the superposition principle, to provide the opportunity for growth toward a mature understanding of this important principle.

### *Electrostatics (Chapters 1 and 2)*

Chapters 1 and 2 on electrostatics deal with the fundamental Coulomb interaction between point charges and with the electric structure of matter, with heavy emphasis on polarization phenomena. We put particular emphasis on the very different kinds of polarization mechanisms in insulators, metals, and ionic solutions (especially the salt solution on the skin). This builds on what the students already know about atoms from chemistry courses and the general culture, but with the recognition that students initially know very little about solids, especially metals.

Students usually come to the course with some notion about "likes repel and unlikes attract." However, most of them have never observed electric repulsion for themselves, and one of the purposes of the desktop experiments involving invisible tape is to acquaint them personally with the basic phenomena. In our experience, most students are deeply puzzled by their observations that show that all ordinary uncharged matter interacts with both plus and minus charges. Chapter 1 deliberately plays up this puzzle to get the students curious about the subject, before we resolve the puzzle in Chapter 2.

We feel that homework problem HW1-1 is exceptionally valuable, in which students are asked to design and carry out an experiment to determine the amount of excess charge on a charged piece of invisible tape. This problem immediately plunges students into real science, including the need for and the value of making approximations. Another good aspect of this project is the fact that there are several different legitimate experimental designs. The analysis also provides a good review of basic mechanics and alerts the students that they will be expected to use what they learned in mechanics in this E&M course.

Another very useful activity at this point is homework problem HW1-4, which involves the use of the computer program *Electric Field Hockey*. The activities and questions in problem HW1-4 reveal important aspects of a $1/d^2$ force law, and they also review important aspects of force and acceleration that are not always clear in the minds of students.

A climax toward which the first two chapters build is the homework problem at the end of Chapter 2 (HW2-1) dealing with getting a spark from a doorknob after shuffling across a carpet. This problem is unusually rich in fundamental physics, with some interesting chemistry as well. It is an excellent topic for classroom discussion to think through together the physics and chemistry of what happens as positive air molecules hit a negative finger (chlorine may be liberated!), or electrons hit a positive finger (liberating hydrogen!). One important point that can emerge from a class discussion is that although in some cases we cannot unequivocally identify a single mechanism for a particular step in the process on the basis of what we already know, we can absolutely rule out various possibilities which may be proposed. It is worth pointing out that because the students typically have little feel for the nature of solids, they may initially think that *nuclei* jump out of the doorknob into the air.

In the absence of Gauss's law, in Chapter 2 we can only assert without proof that excess charge is always found on a surface of a metal, never in the interior. We make it very clear that this is an assertion that we cannot prove at this time, and tell the students that in chapter 10 (on Gauss's law) we will finally be able to show this definitively.

## *Electric field (Chapters 3 and 4)*

In Chapter 3, electric field is introduced with great attention paid to the conceptual difficulties students normally have with this central but difficult concept. The separate reality of electric field is established by invoking relativistic retardation effects, which in our experience are quite meaningful to today's students even if they have never formally studied relativity. We return to the topic of retardation effects several times during the course.

A striking result in Chapter 3 that it is not often discussed is the very steep $1/d^5$ dependence of the force that a point charge exerts on a small piece of neutral matter. This dependence is so steep that at short distances the attraction may swamp the repulsion between two charged objects. It is sometimes possible to observe two charged tapes repel each other at long distances and attract each other when in contact.

Chapter 4 is one of the most mathematically sophisticated chapters in the book, because we attempt to teach the physical meaning of an integral over a charge distribution, and how to set up such integrals. The problem we face is that students often do not really understand that an integral is just a sum but think of it only as an area under some curve. While they typically have little difficulty carrying out a simple given integral, they have little facility with setting up an integral corresponding to a physical situation. Student performance on exam problems indicates that this chapter is largely successful in its goal of teaching students how to set up physical integrals. We think that this success owes quite a bit to the discussion at the beginning of the chapter on numerical integration, and to the homework problem HW4-3 involving a simple numerical integration.

An important theme in Chapters 3 and 4 is the different dependence of electric field on distance near a point charge, a uniform spherical shell, a dipole, a uniform rod, a uniform ring, a uniform disk, and a capacitor. There are two associated pedagogical issues. First, it is easy for students to over-generalize Coulomb's law and think casually that electric field is always proportional to $1/d^2$.

It is important to give lots of practice with various charge distributions to break down this over-generalization.

A second issue is that students need a lot of practice and experience to come to see the value of approximations and the need for care in assessing their range of validity. For example, it is important to know that the field near the center of a uniform disk is nearly uniform but does fall off very slowly with distance near the disk (which leads to there being a fringe field outside the center of a capacitor), and that far from the disk the field is like that of a point charge.

Students typically hold the intuitively appealing but incorrect notion that breakdown in air is due to direct field ionization of air molecules, rather than due to the impact of accelerated free electrons. It is extremely illuminating at the end of Chapter 4 to ask the students to estimate the field strength necessary to yank an outer electron out of an atom (see problem HW4-10). After some thought, students usually are able to come up with the notion that the required field strength is about as large as the field exerted by the rest of the atom on an outer electron, which they themselves calculate to be many orders of magnitude larger than the observed field strength of $3\times10^6$ N/C which is sufficient to trigger a breakdown in air (by accelerating the *free* electrons in the air). This provides an excellent opportunity to discuss the power of science to rule out certain mechanisms, even if science may not be able to *prove* what the real mechanism is.

If a Van de Graaff generator is available, it is advantageous to use it in some electrostatics demonstrations, not only for its value in electrostatics but because a similar device is invoked as a "mechanical battery" in the discussions of circuits in Chapter 6.

We do not introduce Gauss's law at this time, in order to focus on a minimal set of principles with fewer distractions, and to wait until the students have considerably more sophistication about charges and fields. Initially some students tend to confuse "charge" and "field" in speech and writing, and apparently also in thinking. In the absence of Gauss's law, we describe as a plausible but unproved assertion (at the end of Chapter 4) the field due to a uniformly-charged spherical shell, and in an optional section at the end of Chapter 4 we outline a brute-force integration for a spherical shell.

### *Electric current (Chapter 5)*

Chapter 5 opens with a brief discussion of the electric field in a metal wire, and the electron current in the wire as the number of electrons passing some location per unit time. The main body of Chapter 5 is concerned with familiarizing students with the basic phenomena of electric currents in circuits and with using a magnetic compass to detect and measure the presence of electric currents. A major goal of the chapter is to confront the student with a number of deep puzzles about circuit behavior, to stimulate interest in the explanations discussed in later chapters.

Despite its low accuracy, the magnetic compass has a significant pedagogical advantage over a typical ammeter because it is non-invasive, whereas inserting an ammeter necessarily disturbs the circuit. The practical experience of working with the magnetic compass pays additional dividends later when we take up the study of magnetism (Chapter 11). Because of an emphasis on electron current in the metal wires, we introduce a left-hand rule for the magnetic field around a current-carrying wire. Later, when we introduce conventional current in Chapter 9, we change to a right-hand rule. This seems to cause no difficulty for students and may in fact help them make some important physical distinctions.

The most important topic in Chapter 5 is the crucial experiment on current conservation in a series circuit. It is vitally important that this experiment be carried out and discussed carefully. It may be worthwhile to repeat the measurement in class with accurate ammeters. (We have a formal DC circuit lab, included in this manual, in which after completing Chapter 9 students use digital multimeters to make accurate measurements of their circuits.) Students quite naturally believe that "electricity" is used up in a bulb to produce the light, and the problem is with identifying the vague term "electricity" with the flowing electrons. It definitely helps to discuss with the students what would happen if fewer electrons exited the bulb than entered (the bulb continually gets more and more negatively charged, which would have the effect of repelling the oncoming electrons). This argument in terms of dynamics not only helps with understanding current conservation but foreshadows the feedback mechanism examined in Chapter 6.

We derive the important formula for electron current, $i = nAv$. In our experience it is a great help to the students to think in terms of electron current in electrons per second, because it gives the students a mental model and a vocabulary for discussing currents that is much easier to reason with than leaving current as some kind of vague macroscopic flow of "stuff." Moreover, thinking about current as a countable number of elementary particles per second makes a close connection between current conservation and charge conservation.

We also discuss the simple free-electron model for metals and introduce the mobility u, with the drift speed $v = uE$. It is highly useful that the students do *not* recognize this as a microscopic version of Ohm's law. In our experience, students who have previously studied electric circuits have almost always over-generalized Ohm's law, confusing it with the Kirchhoff loop rule. They tend to think that $V = IR$ is a fundamental physical principle rather than merely describing the approximate behavior of some materials (if the temperature doesn't change much). As a result, students often mis-use Ohm's law, for example concluding that there is no potential difference across an open switch, because "V = IR, and there is no I, so there can't be any V. " This has led us to emphasize a very different, microscopic view of current in circuits, and the role of charge and electric field in circuits.

## *Circuits in terms of electric field (Chapters 6 and 7)*

Chapter 5 deliberately leaves the students with a deep mystery about circuits: *Where are the charges that are the sources of the electric field we find inside the wire?* Chapter 6 solves this mystery by showing how a dynamic feedback mechanism leads to the establishment of a steady-state surface-charge distribution that produces that pattern of electric field which satisfies the constraints of current conservation and energy conservation. This novel approach to circuits provides a detailed mechanism for understanding why circuits behave as they do, and it unifies circuit phenomena with electrostatic phenomena. In fact, the same feedback mechanism operates in electro-"static" situations to polarize metal objects, the only difference being that the ultimate result is a final state of static equilibrium (and zero electric field inside a metal), rather than a steady state of steady current (and nonzero electric field inside a metal).

This approach to circuit phenomena involves not only the role of charges and fields but also brings in the structure of matter (especially metals), and conservation of charge, current, and energy. We value the study of circuits for this integration of several different fundamental principles. Our goal is not to make the students into skilled electronics technicians but to provide them with the opportunity to understand circuits at a very basic, atomic level. We hope that this will provide them with a solid foundation for later, more technical studies of circuits.

We deliberately do not use potential in Chapters 6 and 7 in order not to distract the students from the basic issues. This is an example of our minimalist approach to pedagogy: We introduce only just enough concepts and no more in order to be able to analyze phenomena. The concept of electric potential is quite difficult for students, at least if one goes beyond mere formula manipulation. Students find potential to be quite abstract, and indeed it can be considered to be two levels removed from the charges that give rise to it. Analysis in terms of field is closer to the fundamental Coulomb interaction, and this analysis of circuits gives important additional practice with the field concept, an opportunity that is lost if circuits are analyzed solely in terms of potential. Moreover, delaying the introduction of potential has the benefit that students have a more mature understanding of charge and field at the time that they begin the study of potential.

At the college level, many students have previously encountered simple applications of potential and Ohm's law in high-school physics courses. Typically this prior knowledge is not very deep. Students make little distinction between the approximate and phenomenological description of some materials embodied in Ohm's law and the fundamental loop rule which expresses the conservation of energy on a per-charge basis.

We observed an extreme case of a student's attempt to apply Ohm's law inappropriately. The student agreed that the current entering and leaving a light bulb had to be the same, but the current in the bulb filament had to be smaller, since "V = IR, and there is more R in the bulb filament than in the connecting wires, so there has to be less I there." It seems clear that the formula V = IR for many students is a mere algebraic relationship among three independent variables, with no sense of physical mechanism.

In order to understand circuits at a more fundamental level, it is important in Chapters 6 and 7 to avoid potential and Ohm's law entirely. Fortunately, this is easy to do, because the atomic-level analysis based on the fundamental concepts of charge and electric field in Chapter 6 is so utterly unfamiliar to students that they find little incentive to attempt to apply the imperfectly understood concepts of potential and Ohm's law.

Chapter 7 applies the same basic analysis to capacitor circuits. The analysis of RC circuits directly in terms of charges and electric field emphasizes the role of the fringe field just outside the center of the capacitor. If you do cover Chapter 7, it is essential that the students have the half-farad capacitors in their kits, at least during the time devoted to this chapter.

### *Potential (Chapter 8)*

Our treatment of potential emphasizes the connection between field and potential. We take the point of view that field is more fundamental than potential, and that potential difference is *defined* as a path integral of the field. We recognize that in quantum mechanics, potential may be considered more fundamental than field, but in an introductory course it makes sense to emphasis charge and field as fundamental.

We carefully present a four-step procedure for finding potential difference as a path integral (page 285), similar to the procedure in Chapter 4 for finding the electric field of a charge distribution. Examples of potential difference are drawn freely from circuits, in a way that reinforces the connections among charge, field, and potential difference.

The weakening of the average field inside an insulator (the final topic in Chapter 8, which can considered to be optional) is justified by a rather sophisticated argument dealing with path independence of potential. In many textbooks the weakening of the average field is obtained by viewing a polarized insulator not as a collection of polarized molecules but in terms of mathematically-equivalent "surface" charge. We have avoided the concept of equivalent surface charge on insulators because in our experience students easily fall into thinking of these mathematical surface charges as real and, much more seriously, as mobile. This is in large part because a drawing showing only the equivalent surface charge looks dangerously like a polarized metal. Students may easily make the mistake of thinking that an insulator can be discharged by touching it at one location.

In our experience it is very confusing to introductory-level students to talk about the 1/r potential near a point charge, because it undercuts the important notion that it is *differences* of potential that matter. We have therefore been careful in Chapter 8 to talk not about V near a point charge but rather the potential difference $V - V_\infty$.

### *Potential applied to circuits (Chapter 9)*

Chapter 9 revisits the analysis of circuits, now in terms of potential and a macroscopic description that connects to the everyday world of voltmeters and ammeters. The treatment is fairly traditional, but in keeping with our goal of fundamental understanding we omit or downplay some topics that are important in technical electronics, but less important in understanding physics, such as the concept of equivalent resistance. The chapter begins by defining conventional current, which not only involves positive carriers but also is measured in coulombs per second (rather than electrons per second). This implies a shift from a left-hand rule for compass deflections due to electron currents to a right-hand rule for conventional current. This shift seems to cause little or no difficulty for students.

One difficulty that can arise at this point is that students who have not previously been exposed to the standard macroscopic treatment of circuits can feel overwhelmed by the new variables and relations introduced in this chapter. Our solution is to keep emphasizing the macroscopic-microscopic connection, and to remind students that they already understand the basic physics, and the derivation of macroscopic quantities from the previous microscopic view.

In order to reinforce our emphasis on fundamental principles, the Kirchhoff loop and node rules are here referred to as energy-conservation and current-conservation equations. Also, we carefully distinguish between potential difference and emf (non-Coulomb work per unit charge).

A significant application deals with the nature of ammeters and voltmeters. Our goal is to nail down the important facts that an ammeter has very low resistance, and a voltmeter is just a sensitive ammeter in series with a large resistance. (Actually, *digital* multimeters work the opposite way: The fundamental measurement is a voltage, and a digital ammeter is a voltmeter with a low-resistance resistor in parallel!)

At this point in the course we have our students carry out a formal laboratory exercise using modern digital multimeters to observe the loop and node equations in the context of their own circuit kits (batteries, bulbs, nichrome wire, capacitor), plus 47-ohm and 100-ohm ohmic resistors. The lab report is included in this manual.

## *Gauss's law (Chapter 10)*

Now that the students have a rather mature understanding of the relationship between charge and field, we take up the study of Gauss's law. A major difficulty for students is that it is often not quite clear to them what is being related to what. In order to mitigate this confusion, the chapter starts by getting the student to establish a qualitative connection between the patterning of electric field on a closed surface and the nature of the charge inside that closed surface. The next task is to quantify this patterning, which leads in natural steps to a tentative quantitative definition of patterning (electric flux).

Only after establishing in a qualitative way what we are trying to do, and motivating the definition of electric flux, do we proceed with the formal proof of Gauss's law. We then draw on the students' extensive knowledge of the pattern of field near various kinds of charge distributions to check Gauss's law. Finally we are able to find the electric field of a uniform spherical shell and prove that excess charge is only on the surface of a metal, two assertions that were made early in the course.

Another factor that makes Gauss's law more approachable when done late in the course is that by this time the students have had significant practice with integrals, both integrals over charge distributions to determine electric field, and path integrals of electric field to define potential difference. These integrals are probably conceptually easier than the surface integrals encountered in Gauss's law. Also, we write the flux integrand as $\vec{E} \bullet \hat{n}dA$ rather than as $\vec{E} \bullet d\vec{A}$, because in our experience ascribing vector properties to the area element is a conceptual overload; it is conceptually easier for the student to separate out the vector aspect of the area as a unit vector and leave the area as just a scalar.

We do one other thing to lower the burden on the student. In homework, quiz, and exam problems we tell the student what Gaussian surface to use, rather than asking the student to invent a suitable surface. This still leaves much for the student to do, but it does simplify the task significantly.

## *Magnetic field and force (Chapters 11 and 12)*

Often the magnetic force exerted by an existing magnetic field is discussed before discussing how currents make magnetic fields. Because students using our book have had extensive practical experience with the superposition of the magnetic field of the earth and the magnetic field around a current-carrying wire, we have found it appropriate to study the sources of magnetic field (Chapter 11) before studying the effects of magnetic field (Chapter 12).

In Chapter 11 we begin with the tentative operational definition that magnetic field is "whatever it is that is detected by a magnetic compass." Given this detector, plus the superposition principle for magnetic field, the student is able to study in some detail the magnetic field of a magnet, of a straight wire, and of a coil. One of the significant desktop experiments establishes that a bar magnet makes a field that falls off like $1/d^3$, just like the electric-field dependence of an electric dipole.

We emphasize at all times a right-hand rule for cross products rather than components, because we think this is more physical. We have observed that those students who cling to components, unit vectors, and determinants, as emphasized in some previous course, typically demonstrate significantly less physical understanding of the vector character of magnetic fields and forces. The value of the particular right-hand rule that we describe is that it contains an operational definition of the angle whose sine appears in the magnitude, while other right-hand rules only give the direction, and appear to imply that the angle between the two vectors must be a right angle.

We apply the Biot-Savart law to find the magnetic field of a straight wire and of a loop, using a four-step procedure that exactly parallels the procedure introduced in Chapter 4 for finding electric field. The chapter continually shifts back and forth between theory and experiment. In our own course we have the students do the main desktop experiments in class, because we believe that it is very important that they see the phenomena in detail.

Chapter 11 includes a detailed discussion of ferromagnetic materials, with a number of quite satisfying desktop experiments. The question of how a magnet works is one that students are usually quite curious about, so this is an opportunity to engage their attention. We point out briefly that a change of reference frame mixes electric and magnetic fields, so that there must be some deep connection between these two kinds of fields, and we return to this matter in Chapter 12.

In Chapter 11 we break with tradition by not introducing Ampere's law. The magnetic field of a straight wire, a loop, and a solenoid can be obtained in a straight-forward (if calculationally demanding) way from the Biot-Savart law. Of the standard current configurations to which Ampere's law is usually applied, only a toroid really requires Ampere's law. To minimize the number of new concepts, we omit Ampere's law at this time and leave it to Chapter 14, where it is needed in its extended form to complete Maxwell's equations.

In Chapter 12 we study the forces exerted by magnetic fields on moving charges and on current-carrying wires (where we point out that electric forces must also come into play), including torques exerted on coils and other magnetic moments, leading at the end of the chapter to a prediction and measurement of the force between two bar magnets. There is extensive discussion of the Hall effect, because it provides an excellent opportunity to pull together electric and magnetic fields and forces, and an atomic picture of current. Of particular interest is a calculation of the electric and magnetic forces between two protons, which leads to the conclusion that time must run at different rates in different reference frames.

### *Magnetic induction (Chapter 13)*

The treatment of magnetic induction in Chapter 13 puts heavy emphasis on the non-Coulomb electric field associated with a time-varying magnetic field. The emf, the non-Coulomb work per unit charge, is the integral of this non-Coulomb field, which provides an additional opportunity to distinguish between emf and potential difference.

We carefully distinguish among three different influences on an electron in a wire:

1) Coulomb electric field due to surface charges;

2) non-Coulomb electric field due to dB/dt; and

3) magnetic forces associated with changes in the path (motional emf).

The latter two effects are both encompassed within Faraday's law.

Among the applications that are treated in Chapter 13 are electric generators, superconductors, and self-inductance (and its role in RL and LC circuits). At the end of Chapter 13 we offer as optional material some examples of peculiar circuits in which the three different effects contribute in various combinations.

Because induction effects are intrinsically small, they are difficult to observe with desktop experiments. We have our students do a formal lab involving among other things a coil driven sinusoidally and a second coil whose induced voltage is observed on an oscilloscope. This lab gives significant experience with some key aspects of magnetic induction and with RL and LC circuits. The lab report is included in this manual.

To reduce the amount of material in the course to a manageable level, we do not address the technical aspects of AC circuits, with phasor diagrams, but in Chapter 13 and in the associated formal lab we do give the students a strong foundation in the fundamental issues underlying the behavior of AC circuits.

### *Electromagnetic radiation (Chapter 14)*

In Chapter 14 we introduce Ampere's law and its extension by Maxwell, which completes the set of Maxwell's equations. We show that a moving slab of crossed electric and magnetic fields is consistent with Maxwell's equations if the slab moves at a speed c. We stress the effects that electromagnetic radiation has on matter. Based on a quantitative argument by Purcell (E. M. Purcell, *Electricity and Magnetism*, McGraw-Hill 1985, Appendix B), we use a qualitative visual argument for the production of electromagnetic radiation by an accelerated charge, depending heavily on the many previous references in the course to retardation effects. These classical theories of the production of electromagnetic radiation and its effects on matter are then applied to a number of interesting physical problems, including why the sky is blue (and polarized).

It is possible to skip the section on Ampere's law and begin with Maxwell's equations simply as experimentally verified laws. One can also skip the next section, the detailed analysis of the moving slab of radiation, and take as given that a consequence of Maxwell's equations is moving E and B fields (and the transverse electric field is made plausible by the discussion of the electric field of an accelerated charge). Then one can cover just the interesting applications to the effects that radiative fields have on matter.

## How we run our own course

In later chapters we provide detailed information about the course that we teach, using our book. Obviously any teacher will make changes, and we ourselves are unlikely to do things exactly the same from one term to another. But given the radical nature of this new course, we thought it would be helpful to those interested in teaching in this new way to have a very concrete example of one way of managing the course. We provide an example of the course description and weekly assignments given to our students, a daily log of classroom activities during our most recent offering of the course, quizzes, exams, and final exams for two different semesters, two formal lab write-ups, and solutions to selected homework problems.

In our most recent course offering most of the work in the book, except for the desktop experiments, was done outside of scheduled classes. Lectures (for 50 minutes, three times per week) hit the high points of a section of the book but deliberately did not attempt to go over all the details. The lectures were given in an informal, interactive style. Sometimes the lecturer stopped while students worked individually or in small groups through an exercise or experiment in the book, after which there was general discussion of this exercise. Students had the specific assignment outside class of going back over the new sections of the book after the lecture to flesh out the details and get more practice in working with the new concepts.

Recitation sections (20-25 students, for 50 minutes, twice a week) were mainly of two kinds. One of the recitation classes came at the end of the study of a chapter. Students turned in assigned major homework problems at the beginning of the period, and then the teacher guided a discussion of working through the assigned problems. Typically two of the assigned review questions and two of the assigned homework problems were graded; with the human resources available, not all of the assigned review questions and problems could be graded. Detailed solutions were posted the day after problems were turned in.

The other recitation class came earlier, after two or three lectures on a chapter, and this class had a somewhat unusual format. Students worked in groups of two or three for about 15 minutes on selected review questions found at the end of each chapter. This was followed by general discussion of these review questions. The process was repeated, either with additional review questions or with a relatively easy homework problem. Students need not have made special preparations for this interactive problem-solving session other than keeping up on the work. Credit was given for attendance and participation.

When there were critically important desktop experiments to be performed, the schedule was arranged so that the most important experiments were done in class, either in lecture or recitation. Our experience is that students often fail to do some of the critical experiments at home. Moreover, doing the experiments in small groups in class, with recording of group results on the blackboard and discussion of the various group results, seemed to us to have important benefits for the students. It involves them in a social situation of comparison of experimental results that has some of the characteristics of comparisons of experimental results in the world of scientific research. The enterprise was both enjoyable and educational.

In our setting, the three 50-minute lectures and two 50-minute recitation classes every week constitute all of the class time. There is no regular lab period, but twice (for circuits and for magnetic induction) we have the students come to a two-hour formal lab for which they sign up. If your course has a regular lab period, you might choose to retain the lab experiments you have used in the past, or use the time to go more deeply into areas of interest to you and your students.

Some colleagues have run this new course in a workshop mode, where during most of the scheduled class time the students work through the book in small groups, and the teacher wanders around talking with groups individually, with occasional discussion with the class as a whole. While we do some of this ourselves, our impression is that students prefer more variety in the use of class time. It can be particularly wearing if a three-hour lab period is always given over to intense working through the book, unrelieved by other activities.

## *Spending less than a full semester on E&M*

We have been teaching a 15-week semester version of the course (with five 50-minute classes per week and no lab period), as the third semester in a three-semester introductory physics sequence. *If you have more contact hours per week than we have, you may be able to cover all the topics in our book in significantly less than the 15 weeks we require.*

We expect in the future to teach E&M in a 12-week portion of a semester, in a new two-semester introductory physics sequence. The remaining three weeks would be devoted to basic properties of waves (interference, etc.) and would build upon the second half of Chapter 14 on electromagnetic waves and their interactions with matter. The discussion of waves would benefit greatly from the strong emphasis throughout the E&M material on the superposition principle for fields, which should provide a strong sense of mechanism for understanding interference phenomena, especially in optics.

In a 12-week treatment of E&M we intend to completely omit some topics rather than water down those topics that we do treat. We expect to completely omit Chapter 7 (RC circuits analyzed in terms of electric field), Chapter 10 (Gauss's law), and the initial portion of Chapter 14 (sections 14.1-14.3.4). These omissions, while regrettable, need not compromise the other topics, given the structure of our book. These omissions correspond to about 8 days of instruction.

We currently spend 8 class periods on the basic electrostatics of Chapter 1 and Chapter 2, and both we and the students feel that this could be shortened somewhat, with the course getting up to speed a bit quicker. Probably 2 class periods could be harmlessly deleted, if done carefully.

In Chapter 4 (electric field of distributed charges) we would lower our expectations: Instead of attempting to making the students themselves competent in finding the electric field of an arbitrary charge distribution, we would only expect the students to be passively familiar with the integration method and focus mainly just on the patterns of field around a rod, disk, capacitor, and sphere. This lowered requirement is also appropriate given that the students on average will be one semester less far along in the calculus they are taking concurrently (because of the change from a 3-semester physics sequence to a 2-semester sequence). This could save 2 class periods.

In Chapter 8 (potential) it makes sense to omit the topic of the potential of a point charge relative to infinity, and save about 1 class period.

In Chapter 12 (magnetic forces) we would omit sections 12.5-12.5.3 dealing with the magnetic forces and torques on magnetic moments. This would save about 1 class period.

In Chapter 13 (magnetic induction) we would omit self-inductance along with RL and LC circuits, except that some of this material would be retained in the formal lab on magnetic induction. This would save 1 class period.

The total savings come to 15 class periods, or three weeks for us. It is important to keep in mind that we do not have *any* weekly scheduled lab period. If you do, it is quite possible that you can cover our entire book in significantly less than 15 weeks, depending on how you choose to use the lab period.

## Grading qualitative questions

If qualitative physical reasoning is to be taken seriously by the students, there must be qualitative as well as quantitative questions in homework, quizzes, and exams. We are often asked whether it is feasible to grade qualitative exam questions, since reading essays can be very time-consuming. *It is our experience that the grading of qualitative physics problems need not take more time than grading traditional quantitative problems.* (This is assuming that quantitative problems are as usual not graded merely on the basis of the final answer but are analyzed for propagation of errors, and partial credit is given.)

Some care and experience are required to grade qualitative questions efficiently. We have found that it is important not only for efficiency but also for fairness to have a detailed check list of elements that should be present in the student's qualitative explanation, with points assigned for each element. Grading can then mainly be an objective and rapid matter of checking off items on the list rather than a slow and unreliable attempt at a global, impressionistic judgement. The reason that it is possible to grade based on a check list is that there is usually only one, or sometimes two, valid explanations of a physics phenomenon within the constraints of introductory physics. Of course, very occasionally a student may strike out on a highly original path, but there is always time to handle the rare, idiosyncratic case.

One important issue is the use of diagrams. It is very often the case that a qualitative explanation need not contain much prose, if there is an appropriate diagram. At the end of Chapter 2 in the textbook we discuss some aspects of good physics diagrams. It is important to encourage students to use diagrams as tools. Moreover, the more they substitute clear and cogent diagrams for voluminous and inefficient prose, the easier it is to grade their explanations.

We repeat that in our experience we have found that grading qualitative questions need not be more time-consuming or subjective than grading standard quantitative questions. However, one spectacular exception proves the rule that it is important to make a detailed grading key. On one exam we were dismayed to find that it was taking 10 to 15 minutes per student to grade just one qualitative question. We then changed the grading key to make it into a much more detailed list of required elements, whereupon it then took just one or two minutes to grade a student's question!

Here is the question (from exam 2 found in Chapter 6 of this manual):

> **Problem 3 (20 pts):** A capacitor with a slab of glass between the plates is connected to a battery by nichrome wires and allowed to charge completely. Then the slab of glass is removed. Describe and explain what happens. If you give a direction for a current, state whether you are describing *electron current* or *conventional current*.

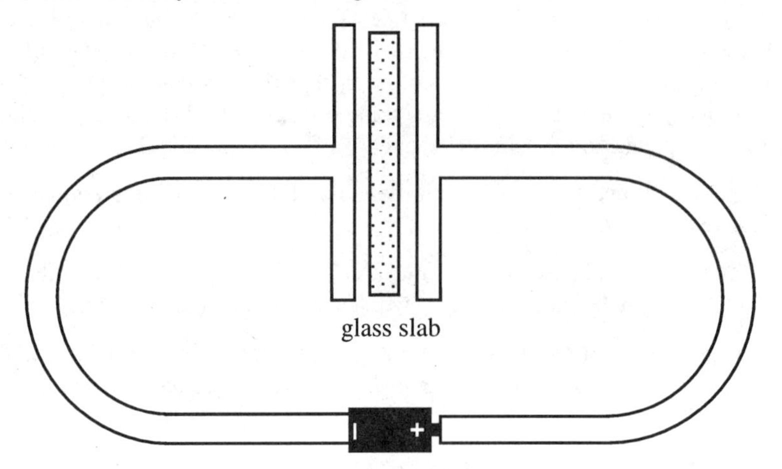


Many explanations were based on charge and electric field. The contribution made by the glass plate to the fringe field in the wire just outside the capacitor goes away when the glass is removed, so that there is a nonzero electric field in the wire in a direction that drives current back through the battery, decreasing the charge on the capacitor plates.

Here is the initial, not detailed grading key that resulted in requiring 10-15 minutes for grading one student (moreover, it appeared that this grading scheme resulted in unreliable grading, in the sense that regrading often gave a very different score):

6 pts initial electric field situation
5 pts electric field situation just after removing glass
5 pts direction of current
4 pts final charge on plates is less than before

There was a similar grading key for those explanations that were based on the use of potential.

Here is the much more detailed grading key for the charge and field explanation that resulted in requiring only 1 to 2 minutes to grade one student (moreover, it appeared that this more efficient grading scheme was also much more reliable and reproducible):

Initial state (10 pts)

1 pt charge on capacitor plates and surface charges on wires
1 pt polarization of glass
2 pts $\vec{E}_{net} = 0$ inside wire in static equilibrium
6 pts $\vec{E}_{net} = \vec{E}_{plates}(1\text{ pt}) + \vec{E}_{glass}(1\text{ pt}) + \vec{E}_{other}(4\text{ pts})$ at specific location inside wire, with directions of each vector

Remove plate (7 pts)

3 pts $\vec{E}_{glass} = 0$
2 pts $\vec{E}_{net}$ direction is to the right inside wires, just outside capacitor
2 pts So electrons flow counter-clockwise

Final state (3 pts)

2 pts $\vec{E}_{net} = 0$ again inside wire in static equilibrium
1 pt Q on plates *less* than before glass was removed

With the undetailed grading key, the grader was expected to look at the student's discussion of the initial configuration of electric field and assign between 0 and 6 points for the quality of this discussion. This proved to be very difficult, time-consuming, and unreproducible. With the detailed grading key, the grader could simply check off the presence or absence of the various elements of a valid discussion of the initial configuration of electric field. This was easy, quick, and highly reproducible.

If you have not previously used qualitative questions, you may initially encounter some difficulties in grading them efficiently, but this is definitely feasible with a bit of practice. The effort is worth it. Unless qualitative questions are as much a part of the course as the standard quantitative problems, students will not take qualitative reasoning seriously and will miss an opportunity to learn how to think like a scientist, with real physical understanding.

In the final chapter of this manual we present solutions to selected homework problems, and for some qualitative problems we provide sample grading keys.

## References on surface charge in circuits

Since the role of surface charge in circuits is not very well known in the physics community, we list a number of relevant references:

H. Haertel, "The electric voltage," in *Aspects of understanding electricity: Proceedings of an international conference*, edited by R. Duit, W. Jung, and C. von Rhöneck (IPN/Schmidt & Klaunig, Kiel, Germany, 1985), pp. 353-362.

H. Haertel, *A qualitative approach to electricity*, Report # IRL87-0001 (Institute for Research on Learning Palo Alto, CA, 1987).

M. A. Heald, "Electric fields and charges in elementary circuits," American Journal of Physics, **52**, 522-526 (1984).

O. Jefimenko, "Demonstration of the electric fields of current-carrying conductors," American Journal of Physics, **30**, 19-21 (1962).

O. Jefimenko, *Electricity and Magnetism* (Appleton-Century-Crofts, New York, 1966), pp. 299-304.

A. Marcus, "The electric field associated with a steady current in long cylindrical conductor," American Journal of Physics, **9**, 225-226 (1941).

W. R. Moreau, S. G. Ryan, S. J. Beuzenberg, and R. W. G. Syme, "Charge density in circuits," American Journal of Physics, **53**, 552-553 (1985).

W. R. Moreau, "Charge distributions on DC circuits and Kirchhoff's laws," European Journal of Physics, **10**, 286-290 (1989).

S. Parker, "Electrostatics and current flow," American Journal of Physics, **38**, 720-723 (1970).

W. G. V. Rosser, "Magnitudes of surface charge distributions associated with electric current flow," American Journal of Physics, **38**, 265-266 (1970).

W. G. V. Rosser, "What makes an electric current 'flow'," American Journal of Physics, **31**, 884-885 (1963).

B. A. Sherwood and R. W. Chabay, "Electrical interactions and the atomic structure of matter: adding qualitative reasoning to a calculus-based electricity and magnetism course," in *Learning Electricity and Electronics with Advanced Educational Technology*, NATO ASI Series F, Vol. 115, edited by M. Caillot (Springer-Verlag, Berlin, 1993), pp. 23-35.

A. Sommerfeld, *Electrodynamics* (Academic Press, New York, 1952), 125-130.

A. Walz, "Fields that accompany currents," in *Aspects of understanding electricity: Proceedings of an international conference*, edited by R. Duit, W. Jung, and C. von Rhöneck (IPN/Schmidt & Klaunig, Kiel, Germany, 1985), pp. 403-412.

# CHAPTER 2

# PROCUREMENT INFORMATION FOR DESKTOP EXPERIMENTS

An essential and required component of the course is a desktop experiment kit covering electrostatics, circuits, and magnetism. The textbook places heavy emphasis on desktop experiments which are tightly integrated with the theory.

Desktop experiment kits are conveniently available through the publisher. If, however, you prefer to assemble your own kits, we offer detailed procurement information gleaned from an earlier period when we assembled our own kits.

## Procuring apparatus for the desktop experiment kits

Each student in our course receives an experiment kit containing the following:

1 roll of "invisible" tape
2 D-cell batteries
1 battery holder, permitting the use of either one or two batteries
2 #14 flashlight bulbs
2 #48 flashlight bulbs
2 flashlight bulb sockets
7 clip leads with alligator clips
2 18-inch (45-cm) lengths of nichrome wire, #26 and #30 gauge
1 liquid-filled magnetic compass
1 half-farad capacitor
1 6-foot (about 2-meter) length of hookup wire
1 bar magnet
3 nails
1 takeout Chinese-food box (4 in. by 3.5 in. by 4 in.) to hold the equipment

The total cost is between $15 and $20, and our students are charged a lab fee to cover this cost. In other institutional settings one might choose to loan out the kits with no charge to the students. To hold down costs, we loan half-farad capacitors to the students for the few weeks when these capacitors are used, and we don't include the capacitor in the kit that the student pays for.

### *Invisible tape*

For the electrostatics experiments we use a cheap local drugstore brand (Rite-Aid) of invisible tape. We got 500-inch tape, a half inch wide, on sale for 50 cents, in a standard holder with a cutting edge which makes it easy to tear off pieces of tape.

Upper (U) Rite-Aid tape is charged *negatively*, whereas with Scotch® brand Magic™ Tape a U tape is charged *positively*. It appears that generic brands of invisible tape are better for these experiments than Scotch brand, which discharges rather rapidly. Each lab period typically consumes half a dozen 20-cm (8-in) lengths of tape, so one tape is sufficient for about 10 lab periods. One also needs a small supply of *aluminum foil and thread.*

### *Batteries and battery holders*

We bought good-quality Panasonic batteries in quantity from Allied Electronics for 26 cents each. We bought Keystone double D-cell battery holders for $1.65 each from Allied Electronics (catalog 2176). It is important that the battery holder permit the use of either one battery or two in a circuit.

### *Bulbs*

The #14 round bulbs draw about 0.3 ampere from two D cells in series. These or equivalent bulbs are easy to get from almost any electronic supplier. At Radio Shack they were sold for 50 cents per bulb (in packages of two bulbs). Later we bought them in quantity from Allied Electronics for 37 cents each.

The #48 long bulbs draw about 0.08 ampere from two D cells in series. These high-resistance bulbs are somewhat hard to find. We first bought them from Delta Education (phone 603-880-6520), stock #54-020-5864, package of 10 for $3.10 (plus shipping and small-order surcharge). Later we bought them for 36 cents each from Allied Electronics (catalog 639S-48).

### Bulb sockets

We ordered bulb sockets from Science Kit & Boreal Laboratories, 777 East Park Drive, Tonawanda NY 14160-6781 (phone 716-874-6020 or 213-944-6317). A package of 15 lamp sockets #66742 is $8.50. On one occasion we bought lamp sockets for $0.25 each from Mouser Electronics, 2401 Highway 287 North, Mansfield TX 76063, catalog number 35LH010. These sockets were less satisfactory, because their shorter ears led to occasional shorting between ear and body if the clip lead rotated on the ear to which it was attached.

### Clip leads

Originally we bought some of the mini-alligator clips (the kind with teeth) from Radio Shack (package of 12 for $1.70), but when they couldn't supply enough of them we got the rest from a local electronics distributor for a slightly higher price. The latter were copper and more satisfactory than the steel Radio Shack clips, because the steel clips can interfere with the compass and the copper clips hold better and make better contact. We used braided hookup wire from Radio Shack, in several colors, and attached clips by soldering. Each kit contained six clip leads that were 10 inches long (including the length of the clips) and one longer clip lead that was 18 inches long.

In the second offering of the course we bought already-assembled clip leads in packages of ten 14-inch leads for $1.95 (catalog ALCP) from Jameco Electronics, 1355 Shoreway Road, Belmont CA 94002. These were more satisfactory in that the alligator clip was covered by a plastic sleeve which prevented tangling of the clip leads in the box. A disadvantage is that the wire is only crimped to the clip, and not particularly well, so that occasionally a clip lead may be open or intermittently open underneath the plastic sleeve, where you might not think to look. The clip leads have steel clips, which can interfere with the compass readings. We included seven of these clip leads per kit.

### Nichrome wire

For ohmic resistors we include 18-inch lengths of #26 and #30 nichrome wire, whose cross-sectional areas differ by about a factor of 2.5 (it is difficult to find wires with the desired area ratio of 2). We ordered nichrome wire from VWR Scientific, 4717 Hinckley Industrial Parkway, Cleveland OH 44109 (800-252-1234).

### Magnetic compass

Originally we used inexpensive small air-filled compasses, but they turned out to be very unsatisfactory for our purposes. The needle would often hang up on the pivot, and give erroneous and misleading results which sometimes induced important misconceptions. Moreover, their physical small size made them hard to work with. The most important consideration is that the glass should be near enough to the needle that a wire carrying 0.3 ampere along the top of the compass should give about a 40-degree deflection of the needle. It is unfortunate that most compasses are built with the glass cover rather far from the needle.

It became clear that we needed to use liquid-filled compasses, and we bought Silva liquid-filled compasses from PASCO Scientific, 10101 Foothills Blvd., Roseville CA 95661, phone 916-786-3800. Stock number EM8631 consists of 5 Silva compasses with 5° divisions for $29 (which is $5.80 each). Science Kit & Boreal Laboratories offers Suunto liquid-filled compasses with 2° divisions for $5.65 each (catalog number 66397-01).

### Magnetism

For desktop magnetism experiments we include a 6-foot (about 2-m) length of hookup wire for making coils, 3 (unmagnetized) iron nails, and a small bar magnet, Edmund Scientific #D31,882 ($14 for a package of 10).

### *Capacitor*

The most expensive and exotic component used in the desktop experiments is an 0.47-farad, 5-volt, non-polar capacitor, NEC "Supercap" FA0H474Z, which is sometimes difficult to order, as many electronics distributors do not carry this item. We ordered ours from Marshall Industries in Pittsburgh, phone 412-788-0441 (contact Ed Cooper); the price was $6.45 each. We don't know of a satisfactory substitute that offers high capacity and low "equivalent series resistance." Do *not* buy capacitors with larger ESR (equivalent series resistance), as it takes longer to discharge them by shorting them, which is confusing to the students. Model FA0H474Z has a stated ESR of 3.5 ohms. There is a new model FE0H474Z with even lower ESR of only 1.8 ohms and a stated price of $6.65, but we haven't tried these yet.

Pricing of these capacitors is bewildering. The list price for one or a few capacitors is about $14, and the price drops rapidly with increasing quantity. But independent of quantity, what you pay is what you are able to negotiate: there really isn't a well-established price. At the time of writing this manual, the publisher is working on assuring a low-cost supply of appropriate capacitors.

# CHAPTER 3

# COURSE DESCRIPTION FOR STUDENTS

Here we present the course description that we give to our own students, outlining the goals and mechanics of the course.

# Physics 3 : Electricity and Magnetism

Lecturer: Bruce Sherwood

Office: 3041 Hamburg Hall

Office hours: any time, but call first to make sure I'm available

## Goals of the course

Physics 3 is the third course in the sequence of introductory calculus-based physics courses taken by science and engineering majors. It covers a single field -- classical electricity and magnetism ("E&M"). Electric and magnetic interactions largely determine the atomic structure of matter, and they are the basis for much of modern science and technology: television, electric lights, computers, telephones, electric motors, etc.

> The goal of this course is to have you engage in a process central to science: the attempt to explain in detail a broad range of electric and magnetic phenomena using a small set of powerful fundamental principles.

We will provide you with the opportunity to acquire a good physical understanding of electric and magnetic phenomena, and of the atomic structure of matter. This course places significant emphasis on qualitative physical reasoning as a complement to the more mathematical quantitative aspects. We want to avoid having the physics get lost in the equations.

## Prerequisites

Prerequisites for the course include Physics 1 and Calculus 1, and you should be taking Calculus 2. We will frequently draw on material from Physics 1, especially Newton's second and third laws. You will need to be able to use vectors, to do simple integration, and to understand the meaning of an integral as a summation of infinitesimal elements. You may find it useful at times to review material from earlier courses.

## Lectures and recitations

| | | | |
|---|---|---|---|
| Lecture | MWF 9:30 | Doherty Hall 1212 | Bruce Sherwood |
| Section A | TuTh 8:30 | Scaife Hall 206 | Bruce Sherwood |
| Section D | TuTh 3:30 | Doherty Hall 2200 | Ruth Chabay |

In all of these classes, both lecture and recitations, we hope you will be aggressive in not letting us proceed if points are not clear! This subject is demanding and difficult, and if you are puzzled or confused you can be sure that many others in the class are too. Do ask lots of questions, in both lecture and recitation.

## Labs

There will be desktop experiments accompanying many of the topics. In addition, there will be two sign-up laboratory exercises during the semester: see assignment sheets. During the week preceding the lab, a sign-up sheet will circulate with a choice of times when the lab will be open. The labs are held in Doherty Hall 2306 (by the Physics 3 bulletin board).

## Special equipment

You will receive an experiment kit for which a lab fee will show up on your tuition bill.

## Interactive textbook

We will use the interactive textbook *Electric and Magnetic Interactions*, by Ruth Chabay and Bruce Sherwood (Wiley 1994). A key component of the course is this book, in which you are asked to carry out experiments, to design experiments, to analyze phenomena, to work out small examples, to make some of the steps in derivations, etc.

The high points of new material will be discussed in lecture. After the lecture, it is important that you work through the new material in the book in detail, to help fix the concepts in your mind. It is important that you take this study assignment seriously, a day at a time. If you ignore the book until it is time to attack an assigned homework problem, you may waste a lot of time floundering around, desperately searching for a non-existent magic formula somewhere in the chapter that sort of matches the homework problem, and you will lose the opportunity to acquire a deep understanding of the material.

If on the other hand you devote a modest amount of daily time to working through the new sections of the book, you will be in a position to attack the homework problems efficiently, based on a clear understanding of the fundamental physical principles that underlie the analysis of all the homework problems. You will also be well prepared for quizzes and exams, which test your understanding of fundamental principles rather than your ability to plug numbers into secondary special-case formulas.

## Homework problems

Selected homework problems will be graded and will count toward the final course grade. Solution sets will be posted on the bulletin board outside of Doherty Hall 2306 and will be available in binders on reserve in the Engineering and Science Library. **Homework is due at the start of the specified class; it will receive half credit if handed in by the start of the following class; it will not be graded if handed in later.**

## Quiz every week

Unless explicitly stated otherwise, there will normally be a **short quiz every week** (see assignment sheet). Quizzes are given before we have fully completed a topic and are designed to test your understanding of the most fundamental aspects of the topic, before you attack the more difficult and more comprehensive homework problems.

We will drop your two lowest quizzes (with an unexcused absence counted as zero). Make-up quizzes will not be given, but in the case of an *excused* absence we will average your other quiz scores.

## Exams

**There will be three exams and a 3-hour final exam.** All exams are closed-book, but relevant formulas and constants will be provided where needed. Hour exams are at 6:30 pm to permit you to take more than an hour if you need extra time. If you have an unavoidable problem due to illness or family matters, contact the lecturer no later than the day of the exam. No excuses after the exam date will be accepted. No makeup exams will be given (in the case of an *excused* absence we will average your other exam scores).

## Computer work

From time to time during the semester we may assign certain computer homework to be done outside class. These computer activities are designed to complement your other means for learning the material. The programs use interactive graphics to help you visualize complex concepts and to provide an opportunity to explore the consequences of electrical theory. No special computer knowledge is needed to benefit from these programs. You will be asked to turn in a short report based on your computer work.

## Collaborative work

Scientists and engineers normally work in groups, and social interactions are critical to their work. Most good ideas grow out of discussions with colleagues. ***In this course, we want you to work with others as much as possible.*** Study together, help your partners to get over confusions, ask each other questions, and critique each others' homework and lab write-ups. Teach each other! You can learn a great deal by teaching. But *do list your partners* on any papers you turn in for grading, just as scientists cite collaborators in their papers.

While collaboration is the rule in technical work, evaluations of individuals also play an important role in science and engineering. Quizzes and exams are to be done without help from others.

## Grades

The final grade will be determined on the following basis:

**35%** **final exam**

**35%** **three hour exams**

**15%** **quizzes**

**15%** **other work (homework, formal labs, computer exercises)**

Class participation in recitation may be used to judge borderline cases.

## Help

You should ask lots of questions in lecture and recitation. Each TA will establish office hours, and you are encouraged to take advantage of this help. If you miss work because of illness or other circumstances, please let your TA or the lecturer know as soon as possible. The sooner we know about these situations, the better we will be able to help you make up the work. We will do what we can to help you complete the course satisfactorily, but an "incomplete" grade cannot be given simply because you have fallen behind.

## Extra copies of handouts

Extra copies of handouts are available in the offices of Bruce Sherwood (Hamburg Hall 3041) and Ruth Chabay (Hamburg Hall 3043).

# CHAPTER 4

# SAMPLE ASSIGNMENT SHEETS FOR STUDENTS (SEMESTER A)

In this appendix we present the weekly assignment sheets that we gave to our own students during semester "A" (the next two chapters deal with this same semester). These assignment sheets are very detailed and include assignments of specific sections of the book to work through on a daily basis after the high points have been discussed in lecture.

## Week 1

| | |
|---|---|
| **Main topics:** | • basic aspects of electric forces<br>• Coulomb's law<br>• the superposition principle<br>• conservation of charge |
| **Mon.** | No class |
| **Tues.** | Explanation of course organization<br>Begin Chapter 1, The Interactions of Electric Charges<br>Outside class: re-read syllabus and preface; study sections 1.1-1.4.1 inclusive |
| **Wed.** | Coulomb's law<br>Outside class: study sections 1.4.2-1.7.3 inclusive |
| **Thurs.** | Discuss selected Ch. 1 review questions (p. 28-29), for credit<br>Work with partners to plan attack on problem HW1-1, due Tuesday<br>Outside class: review for quiz; get started on work due Tuesday |
| **Fri.** | Short quiz on basic aspects of Chapter 1, **at start of class: Be prompt!**<br>Begin Chapter 2, Charges in Matter<br>Outside class: study sections 2.1-2.4 inclusive |

## Week 2

| | |
|---|---|
| **Main topics:** | • permanent and induced electric dipoles<br>• polarization of insulators<br>• polarization of conductors: ionic solutions and metals<br>• charging and discharging |
| **Mon.** | Insulators and conductors<br>Outside class: study sections 2.5-2.7.5 inclusive (before Wed.) |
| **Tues.** | At start of class, turn in<br>review questions RQ1-5, RQ1-7, RQ1-8<br>homework problems HW1-1, HW1-2, HW1-3, HW1-6<br>Discussion of homework |
| **Wed.** | Charging and discharging; *Electric Field Hockey* (due Thurs. next week)<br>Outside class: study sections 2.8-2.9.5 inclusive |
| **Thurs.** | Discuss selected Ch. 2 review questions, for credit<br>Discuss homework problem HW2-1 due Tuesday<br>Outside class: review for quiz; get started on work due Tuesday |
| **Fri.** | Short quiz on basic aspects of Chapter 2, **at start of class: Be prompt!**<br>Begin Chapter 3, Electric Field<br>Outside class: study sections 3.1-3.1.7 inclusive |

## Week 3

| | |
|---|---|
| **Main topic:** | • electric field—concept and applications |
| **Mon.** | Superposition principle for electric field; field of a dipole<br>Outside class: study sections 3.2-3.3.1 inclusive (before Wed.) |
| **Tues.** | At start of class, turn in<br>review questions RQ2-3, RQ2-4, RQ2-7<br>homework problems HW2-1, HW2-7, HW2-9<br>Discussion of homework |
| **Wed.** | Applications of electric field<br>Outside class: study sections 3.3.2-3.5 inclusive |
| **Thurs.** | At start of class, turn in homework problem HW1-4 (*Electric Field Hockey*)<br>Discuss selected Ch. 3 review questions, for credit<br>Discuss a Chapter 3 homework problem<br>Outside class: review for quiz; get started on work due Tuesday |
| **Fri.** | Short quiz on basic aspects of Chapter 3, **at start of class: Be prompt!**<br>Conclusion of Ch. 3: limitations of the electric field concept; retardation |

## Week 4

| | |
|---|---|
| **Main topics:** | • how to calculate the electric field of distributed charges<br>• electric field of rod, ring, disk, two disks (capacitor), and spherical shell |
| **Mon.** | Begin Chapter 4, The Electric Field of Distributed Charges<br>Electric field of uniformly-charged rod<br>Outside class: study sections 4.1-4.2.1 inclusive (before Wed.) |
| **Tues.** | At start of class, turn in<br>review questions RQ3-3, RQ3-8, RQ3-9<br>homework problems HW3-5, HW3-6, HW3-8<br>Discussion of homework |
| **Wed.** | Electric field of ring, disk, and capacitor<br>Outside class: study sections 4.3-4.5 inclusive (before Fri.) |
| **Thurs.** | Discuss selected Chapter 4 review questions, for credit<br>Get started on Chapter 4 homework |
| **Fri.** | Electric field of a spherical shell; applications<br>Outside class: study section 4.6 (before Mon.) |

## Week 5

| | |
|---|---|
| **Mon.** | Review for exam |
| **Tues.** | At start of class, turn in<br>review questions RQ4-4, RQ4-7<br>homework problems HW4-3, HW4-5, HW4-13 |
| **Wed.** | No class because there is an exam, but we will be in the lecture room to answer questions. |

**Exam #1, 6:30 pm Wednesday**

**Exam covers Chapter 1 through 4**

| | |
|---|---|
| **Thurs.** | In class, begin circuit experiments (Chapter 5) |
| **Fri.** | Continue Chapter 5 |

## Week 6

<table>
<tr><td colspan="2">

**Main topics:**

- electric field in circuits
- what charges make the electric field in a circuit
- feedback
- energy input and output in a circuit

</td></tr>
<tr><td>Mon.</td><td>Finish Chapter 5: Current conservation; electron current & drift speed; start-stop motion.<br><br>Begin Chapter 6: Electric field in circuits<br><br>Outside class: study sections 5.5-5.5.3 inclusive<br>study sections 6.1-6.3 inclusive (before Tues.)</td></tr>
<tr><td>Tues.</td><td>The initial transient; feedback. Surface charge & resistors.<br><br>Discuss selected review questions, for credit.<br><br>Outside class: study sections 6.4-6.5.3 inclusive (before Wed.)</td></tr>
<tr><td>Wed.</td><td>Review of resistors. Batteries: Energy input and output.<br><br>Outside class: study sections 6.6-6.6.4 inclusive (before Thurs.)</td></tr>
<tr><td>Thurs.</td><td>Applications of the theory. BRING CIRCUIT KITS<br><br>Discuss selected review questions, for credit.<br><br>Outside class: study sections 6.7-6.9 inclusive (before Mon.)</td></tr>
<tr><td>Fri.</td><td>Short quiz on basic aspects of Chapter 6, at start of class: Be prompt!<br><br>Work through some Chapter 6 homework problems.</td></tr>
</table>

## Week 7

**Main topics:**

- capacitors in circuits—non-steady-state currents
- electric potential

**Mon.** **BRING CIRCUIT KITS**

Chapter 7—capacitors in circuits.

Outside class: study sections 7.1-7.5 inclusive (before Wed.)

**Tues.** At start of class, turn in

review questions RQ6-4, RQ6-5, RQ6-7, RQ6-9

homework problems HW6-1, HW6-3, HW6-5

Discussion of homework

**Wed.** **BRING CIRCUIT KITS**

Finish Chapter 7.

Outside class: study sections 7.6-7.7 inclusive

**NOTE SHORT ASSIGNMENT TO TURN IN ON THURSDAY**

**Thurs.** At start of class, turn in

reports on selected sections in Chapter 7 (to be announced)

review question RQ7-4

homework problems HW7-4, HW7-5

**Fri.** Short quiz on basic aspects of Chapter 7, **at start of class: Be prompt!**

Chapter 8—electric potential

Outside class: study sections 8.1-8.2.1 inclusive (before Tues.)

## Week 8

| | |
|---|---|
| **Main topics:** | • electric potential |
| **Mon.** | *No class—midsemester break*<br>Outside class: study sections 8.1-8.2.1 inclusive (before Tues.) |
| **Tues.** | A procedure for calculating potential difference; applications<br>Outside class: study sections 8.2.2-8.4, stopping before 8.4.1 (before Wed.) |
| **Wed.** | Path independence of potential difference<br>Outside class: study sections 8.4.1-8.7.4 inclusive (before Fri.) |
| **Thurs.** | At start of class, turn in<br>easy homework problems HW8-1, HW8-3, HW8-4, HW8-5<br>Discussion of homework |
| **Fri.** | Short quiz on basic aspects of Chapter 8, **at start of class: Be prompt!**<br>Continue Chapter 8<br>Outside class: study sections 8.8-8.8.3 inclusive, skim section 8.10, study section 8.10.1 (before Mon.) |

## Week 9

**SIGN UP FOR CIRCUIT LAB**
**this Fri. (2:30-4:30 or 4:30-6:30)**
**or next Mon. (12:30-2:30 , 2:30-4:30, or 4:30-6:30)**
**-- Bring your kit to the lab! --**
**Lab takes 2 hours, including turning in report at end.**

| | |
|---|---|
| **Main topics:** | • finish up electric potential<br>• electric circuits in terms of potential (macroscopic analysis) |
| **Mon.** | Quiz on Chapter 8<br>Finish Chapter 8—potential of a point charge relative to infinity<br>Outside class: study section 9.1.1 on "conventional current" (before Wed.) |
| **Tues.** | At start of class, turn in<br>homework problems HW8-7, HW8-13, HW8-16, HW8-17<br>Discussion of homework. |
| **Wed.** | Begin Chapter 9: Resistance; emf in a circuit<br>Outside class: study sections 9.1.2-9.2.2 (before Thurs.) |
| **Thurs.** | **Bring your kit to recitation!**<br>Loop and node equations; power; internal resistance.<br>Outside class: study sections 9.3-9.4.1 inclusive (before Fri.) |
| **Fri.** | **Bring your kit to lecture!**<br>Multi-loop circuits. Ammeters and voltmeters.<br>**Circuit Lab (if you signed up for Fri.)—*bring your kit!***<br>Outside class: study sections 9.4.2-9.5 inclusive (before Mon.) |

## Week 10

| | |
|---|---|
| **Main topics:** | • finish up electric circuits in terms of potential (macroscopic analysis) |
| **Mon.** | Finish Chapter 9—demonstration of surface charge in a 10000-volt circuit; RC circuits.<br>**Circuit Lab (if you signed up for Mon.)—*bring your kit!***<br>Outside class: study section 9.6 (before Tues.) |
| **Tues.** | At start of class, turn in<br>review questions RQ9-3, RQ9-7, RQ9-9<br>homework problems HW9-2, HW9-3, HW9-5, HW9-9<br>Discussion of homework. |
| **Wed.** | No class because there is an exam, but we will be in the lecture room to answer questions. |

**Exam #2, 6:30 pm Wednesday**

**Exam covers Chapter 5 through 9**

| | |
|---|---|
| **Thurs.** | Begin Chapter 10 on Gauss's law<br>Outside class: study sections 10.1-10.3.5 inclusive (before Fri.) |
| **Fri.** | Continue Chapter 10<br>Outside class: study sections 10.4-10.5.6 inclusive (before Mon.) |

## Week 11

| | |
|---|---|
| **Main topics:** | • finish Gauss's law<br>• the magnetic field of magnets and of current-carrying wires<br>• the atomic theory of magnets |
| **Mon.** | Begin Chapter 11: Magnetic field. **BRING EXPERIMENT KITS**<br>Outside class: study sections 11.1-11.2.2 inclusive (before Wed.) |
| **Tues.** | At start of class, turn in<br>review questions RQ10-1, RQ10-2, RQ10-3, RQ10-6<br>Do homework problem HW10-2 in class for credit. |
| **Wed.** | Magnetic field of straight wires. **BRING EXPERIMENT KITS**<br>Outside class: study sections 11.2.3-11.4.2 inclusive (before Thurs.) |
| **Thurs.** | Magnetic field of coils. **BRING EXPERIMENT KITS**<br>Outside class: study sections 11.4.3-11.5.2 (before Fri.) |
| **Fri.** | Short quiz on basic aspects of Chapter 11, **at start of class: Be prompt!**<br>Atomic theory of magnets.<br>Outside class: study sections 11.6-11.7.2 (before next class) |

*Spring vacation week*

## Week 12

| | |
|---|---|
| **Main topic:** | • magnetic force |
| **Mon.** | Begin Chapter 12: Magnetic force.<br>Demonstration of magnetic forces on moving charges. Hall effect.<br>Outside class: study sections 12.1-12.2 inclusive (before Wed.) |
| **Tues.** | At start of class, turn in<br>review questions RQ11-2, RQ11-3, RQ11-9<br>homework problems HW11-1, HW11-2, HW11-7 |
| **Wed.** | Relativistic effects. Demonstration of forces on current-carrying wires.<br>Outside class: study sections 12.3-12.4.1 inclusive (before Thurs.) |
| **Thurs.** | Discuss selected Chapter 12 review questions, for credit<br>Work on one of the homework problems for next Tuesday. |
| **Fri.** | Short quiz on basic aspects of Chapter 12, **at start of class: Be prompt!**<br>Forces on magnetic moments. **BRING YOUR BAR MAGNETS!**<br>Outside class: study sections 12.5-12.5.3 inclusive (before Mon.) |

## Week 13

**SIGN UP FOR INDUCTION LAB**

**Next Week**

**Mon. Apr. 18 (12:30-2:30 , 2:30-4:30 or 4:30-6:30)**

**or Tues. Apr. 19 (12:30-2:30 , 2:30-4:30, or 4:30-6:30)**

**-- Bring your experiment kit to the lab,**

**and the value of $\mu$ for your bar magnet --**

**Lab takes 2 hours, including turning in report at end.**

| **Main topic:** | |
|---|---|
| | • magnetic induction—effects of a time-varying magnetic field |
| **Mon.** | Begin Chapter 13: Magnetic induction.<br>Demonstration of magnetic induction phenomena.<br>Outside class: study sections 13.1-13.2.3 inclusive (before Wed.) |
| **Tues.** | At start of class, turn in<br>review questions RQ12-7, RQ12-8<br>homework problems HW12-1, HW12-3, HW12-4, HW12-10<br>results of exercises 12.5.3c-12.5.3e (one magnet picking up another magnet) |
| **Wed.** | Lenz's rule. Applying Faraday's law and Lenz's rule.<br>Outside class: study sections 13.2.4-13.3.4 inclusive (before Thurs.) |
| **Thurs.** | For credit, discuss review question RQ13-1, and homework problems HW13-3 (omit part e) and HW13-6. You do *not* need to work these out before coming to class. |
| **Fri.** | *No class—Spring Carnival.* |

## Week 14

**INDUCTION LAB**

**Mon. (12:30-2:30 , 2:30-4:30 or 4:30-6:30)**

**or Tues. (12:30-2:30 , 2:30-4:30, or 4:30-6:30)**

**-- Bring your experiment kit to the lab,**

**and the value of μ for your bar magnet from exercise 11.5a --**

**Lab takes 2 hours, including turning in report at end.**

| | |
|---|---|
| **Main topics:** | • motional emf<br>• superconductors<br>• self-inductance |
| **Mon.** | Motional emf; electric generators.<br>Demonstration of superconductivity.<br>Outside class: study sections 13.4-13.7.2 inclusive (before Wed.) |
| **Tues.** | At start of class, turn in<br>review questions RQ13-4, RQ13-5<br>homework problems HW13-7, HW13-10 |
| **Wed.** | RL and LC circuits.<br>Outside class: study sections 13.8-13.8.3 inclusive (before Thurs.) |
| **Thurs.** | Discuss selected Chapter 13 review questions and homework, for credit |
| **Fri.** | Review for exam |

## Week 15

| **Main topic:** | |
|---|---|
| | • electromagnetic radiation: how it is produced, and its effects |
| **Mon.** | No class because there is an exam, but we will be in the lecture room to answer questions. |

**Exam #3, 6:30 pm Monday**

**Exam covers Chapter 10 through 13**

| | |
|---|---|
| **Tues.** | Begin Chapter 14: Ampere's law and Maxwell's extension of it. Electromagnetic radiation—interplay of electric and magnetic fields. Effects of electromagnetic radiation on matter.<br>Outside class: study sections 14.1-14.4.4 inclusive (before Wed.) Just skim sections 14.1-14.3.4, but be sure you are clear about the effects discussed in sections 14.4-14.4.4 inclusive. |
| **Wed.** | *Accelerated* charges produce electromagnetic radiation. Demonstration of wireless electromagnetic transmission of energy.<br>Outside class: study sections 14.5-14.7.3 inclusive (before Thurs.) |
| **Thurs.** | Discuss selected Chapter 14 review questions and homework problems, for credit |
| **Fri.** | Short quiz on basic aspects of Chapter 14, **at start of class: Be prompt!**<br>Why the sky is polarized, and blue.<br>Fill out course evaluations. |

**Final exam (3 hours), 5:30 pm Friday**

**You may bring one sheet of 8.5"×11" paper with any notes you like on both sides. In addition, we will provide a formula sheet.**

# CHAPTER 5

# DAILY LOG OF CLASS ACTIVITIES (SEMESTER A)

In this chapter we present a very detailed daily log of class activities. This log was written day by day during the seventh semester that we taught this course (semester "A").

## Structure of a typical week

There were three 50-minute lectures (L1, L2, L3 on Monday, Wednesday, and Friday) and two 50-minute recitation classes (R1, R2 on Tuesday and Thursday) every week. There was no regular lab period, but in addition to the integrated desktop experiments there were two 2-hour sign-up labs during the semester (quantitative circuits and magnetic induction). An entire 15-week semester was devoted to electricity and magnetism.

The cycle of activities for a typical week looked like this (modified of course by exams, vacations, etc.), starting with the Friday lecture (L3), with time in minutes given for each component:

L3 15 Quiz over very basic aspects of the old chapter.

35 High points of the first part of the new chapter.

Weekend—work on homework problems from old chapter, review start of new chapter.

L1 50 High points of the middle of the new chapter.

R1 50 At start of class, students hand in assigned homework problems from old chapter. Then there is a general discussion of these problems.

L2 50 High points of the end of the new chapter.

R2 20 Students work on selected review questions in small groups.

20 Discussion of these review questions.

10 Students work in small groups on an assigned homework problem due next week.

L3 repeat cycle

The quiz during lecture L3 came before students had tackled the homework problems for the old chapter, so the quiz dealt only with the most basic issues and concepts of the old chapter. The finale of a chapter was the class discussion of solutions to the homework problems in recitation class R1. Note that the students could work on the homework problems over the weekend.

The lectures followed the book very closely but only hit the high points and did not attempt to cover everything. Often during the "lecture" students were asked to work out something in the book and then this point was opened to general discussion and amplification. After each lecture, the students had the assignment to go back over that section of the book and complete it in detail.

*One must make it clear to students that the reading assignments are to be taken seriously.* It is often the case in introductory physics courses that students do not really study the body of a textbook chapter but simply flip through it when doing the homework problems, looking for useful secondary formulas. This contributes to a lack of fundamental physical understanding and an emphasis on formula matching. One of the reasons this occurs is that without working in-line exercises such as are in our book, it is very difficult for students to read a physics textbook on their own with understanding. Some students may "read through" our book without doing the exercises because they feel that they don't have time to work out the details. However, this is actually very inefficient, because little or no learning takes place, and the time spent in passive reading is mostly wasted.

We tried to have all of the most critical desktop experiments done during lecture or recitation classes, where there were partners available. Some students are unlikely to do all the critical experiments at home, and doing the experiments in class makes sure that the key phenomena have been observed, compared, and discussed. When experiments were done in class, we recorded on the board the results of each group as they completed the measurement. The table of data from the various groups was a rich source for discussion about the experiment.

**Week 1**

L1 30 Pass out invisible tape, give overview of course (go over syllabus and book preface). Emphasize unique characteristics of this course: goal of explaining a broad range of phenomena with just a few fundamental principles; interactive book (and how to use it); emphasis on qualitative reasoning rather than formula manipulation; emphasis on good scientific explanations.

20 Work on Chapter 1, preferably in groups of two. To forestall possible objections to the simplicity of the equipment, it is useful to explain that these seemingly simple experiments will quickly lead to some very deep physical puzzles, and fancy equipment is irrelevant at this point.

R1 No class (in the first week of the course, there is only one recitation class rather than two).

L2 50 Guided tour of the high points of the second half of Chapter 1. Explain that it is up to the students to go back over this material in the book themselves in detail. The purpose of the lecture is to emphasize the most important points.

Point out to students that usually there are no easy homework problems at the end of the chapter. Instead, there are short in-line exercises which the students should do to learn the material fully.

Section 1.2.9: ask about their observations of the key properties (repulsion, attraction, fast fall-off with distance, force along line between objects).

Exercise 1.3.1a: ask about student observations.

Section 1.3.1: we're left with 3 puzzles.

Summarize Coulomb's law; students do exercises 1.4.3a and 1.4.3c.

Have students do exercises 1.5.1a and 1.5.3a (sign of U tape).

Conservation of charge: discuss exercise 1.6.2a.

Superposition principle: have students attempt observations in section 1.7.3.

R2 10 Students work in groups on selected review questions (RQ1-1, 1-2, 1-4, 1-6).

15 Discuss these selected review questions.

10 Small group discussion, then class discussion of how to put gross upper and lower limits on the amount of charge on a tape (problem HW1-1): one excess electron, or *all* the electrons.

15 Students work in groups of two or three to plan how to attack problem HW1-1.

L3 15 Short quiz on very basic concepts in Chapter 1.

35 Introduce dipoles at beginning of Chapter 2 (section 2.1.1). Students work exercise 2.1.1a, then we discuss it, and go through the detailed calculation leading to the $1/d^3$ behavior. Then ask the question posed by exercise 2.1.1e.

On the board, build up a "multiple exposure" like that shown in section 2.2.1, and discuss its symmetry, which suggests that it would have no effect on a distant charge.

Discuss the "multiple exposure" of the polarized hydrogen atom in section 2.2.3.

Finale: discuss section 2.3.1, why your hand attracts charged tapes. This resolves one of the major puzzles of Chapter 1.

**Week 2**

L1 50 Give each student a small piece of aluminum foil (about 8 cm by 1 cm) for Chapter 2 experiments.

Draw attention to the twist of a dipole that they should have observed in the assigned work.

Students do 2.5.2a—is tape an insulator? Then discuss, emphasizing the point that the charges would push each other onto the uncharged part of the tape if the charges were mobile. Also discuss the fact that the observation rules out mobile charges inside the tape, because internal mobile charges would have moved into the uncharged part of the tape. So charges don't move on or in the tape.

Polarization of an insulator, and how to draw it.

Polarization of an ionic solution: students do 2.7.1a, then we discuss it. Show that the polarization reduces the net force on an ion in the interior of the liquid, and encourage students to think about how far this process will go. Eventually we see together that the net force will go to zero in static equilibrium. (But also pointed out that the ions are jostled around in the liquid, so really we have a concentration gradient, not two nicely defined + and - sheets of charge.) Key point is superposition: the external force is not affected, but the net force goes to zero.

Polarization of a metal. Quantum mechanics is really needed for a full analysis, but fortunately a simple picture takes us a long way. A block of copper is not simply a bunch of copper atoms; rather it is a lattice of ions in a sea of mobile electrons, each on the average free of force. Polarization like ionic solution, but only one kind of mobile charge. Informal version of proof by contradiction to show again that in static equilibrium the net force is zero inside (and no net charge anywhere inside).

Drawing conventions: insulator vs. metal.

Summary of sections 2.7.4 and 2.7.5.

Students do exercises 2.7.5a and 2.7.5b, then we discuss the solutions.

R1 50 Discussion of Chapter 1 homework assignment. At start of class, before handing in homework, students announce their results for the amount of charge on a tape (problem HW1-1), and these values are written on the board (coulombs and electron charges). Students then hand in their homework, and we discuss the homework, with emphasis on problem HW1-1, which is rich in physics principles. Ask students to describe their methods, and display the geometries on the board. Do the force analysis for a tape hanging at an angle.

Carefully discuss what simplifying assumptions had to be made (assumption of point charge, and assumption that $Q_1 = Q_2$, unless they use attraction between U and L tapes made from a neutral sandwich). Some students also recognize that they should keep the tapes away from large neutral objects, since that adds an additional attractive force. Can discuss whether the (gross!) assumption of point charges leads to a lower limit or an upper limit on the actual amount of charge on a tape. Some students hang small pieces of tape from threads, so as to approximate point charges.

Before discussing problem HW1-2 it is good to ask whether their results for the number of electron charges (which may represent the number of ions) is a "large" number. Then discuss problem HW1-2 and show that this seemingly "large" number is in fact an extremely small number, since large and small have physical meaning here only in relation to the far larger number of molecules in the surface.

In a typical class, all students who don't make arithmetic or units errors find about the same amount of charge on a tape. Problem HW1-3 gives a mechanism for why a tape has about the same amount of charge every time you charge it: the breakdown strength of air limits the amount of charge separation attainable. It appears that charge transfer is nearly "maximal" in the sense that it seems to be limited by air breakdown and might well be much larger in vacuum. The agreement of results among different students justifies after the fact the assumption that $Q_1 = Q_2$.

Also discussed briefly the other assigned homework. If there is time left over, one can have the students work in groups on a simple unassigned Chapter 2 homework problem such as HW2-3, then discuss the problem together.

In our case, only two or three of the assigned problems and questions are actually graded for record, and the students don't know which ones will be graded. The homework assignment forces students to wrestle with the problems before coming to the recitation class where we go over the problems together. This makes the discussion quite fruitful.

L2 15 Students do charging experiments (sections 2.8.1-2.8.3).

10 Discussion of charging and discharging with the body (section 2.8.4). It is fun to discuss the chemical species involved in touching positively or negatively charged metal with your finger. For example, if you touch a negatively charged piece of metal, one of the likely mechanisms is that a sodium ion in the salt water layer on your skin combines with an electron in the metal to produce atomic sodium, which can then react with the water to make sodium hydroxide and hydrogen!

Some students observe that after charging the metal foil by contact with a charged pen, the pen does repel the foil, except that at very close distances it actually *attracts* the foil! Can discuss this odd phenomenon as a competition between the repulsion of like-charged objects and the attraction between charged objects and neutral matter. Evidently the attractive force due to induced polarization falls off with distance more rapidly than the $1/d^2$ Coulomb repulsion (see details in Chapter 3, section 3.3.3).

5 Students do charging by induction (section 2.8.5).

5 Discussion of charging by induction.

15 Introduce the program *Electric Field Hockey*. Skip the instructions, go right to Level 1. First place a single charge behind the ball and drive the ball into the sticky area. Then move the propelling charge to drive the ball above the sticky area. Place a second charge near the trajectory and see the ball be driven sharply down into the sticky area. Then pull the second charge up, quite far from the trajectory, and try again; this time the deflection is quite small, and immediately the students see one aspect of the rapid $1/d^2$ fall-off. Then move the second charge until you score a goal. Turn on the force vector and repeat the trajectory.

Give a very quick tour of Levels 2 through 6, just to show the increasing level of challenge, but without solving any of these levels. Alert the students that later they will have the assignment to use the program themselves to do problem HW1-4. The main purpose of this is to get a gut feel for the short-range and long-range aspects of the Coulomb interaction. A secondary purpose is to review the basic mechanics of frictionless motion due to known forces.

R2 10 Point out to students that they should be able to do all the review questions, whether assigned or not. Students work in groups on selected review questions (RQ2-5, 2-9, 2-10, 2-12). If extra time, also do RQ2-2.

15 Discuss these selected review questions.

15 Give an overview of electric discharges in gas as in the footnote to problem HW2-1, but talking only about a charged metal sphere near a neutral metal sphere, leaving the rest of the problem concerning the body as a student assignment. All of this can be the basis of excellent discussions, because it involves a lot of basic physics and chemistry.

10 Students discuss problem HW2-1 in groups of two or three. The problem will be handed in next week. Point out that it is the quality of the explanation that counts, and draw their attention to section 2.9 on how to write a good scientific explanation.

L3 15 Short quiz on very basic concepts in Chapter 2.

10 Demonstrate a Van de Graaff generator. Explain the basic operation of charge transport on the conveyor belt. Make sparks by bringing a grounded conductor near the charged sphere, reviewing the nature of spark formation (problem HW2-1). Charge by contact a metal sphere on an insulating stand. Test with invisible tape to establish sign of charge by repulsion; show that when tape gets very close it actually sticks to the sphere! (The $1/d^5$ attraction wins out over the $1/d^2$ repulsion; details later in Chapter 3.) It is useful to demonstrate the Van de Graaff generator at some point before reaching Chapter 6, where we introduce a "mechanical battery" that is a similar device.

20 Introduce the concept of electric field, paralleling the treatment at the beginning of Chapter 3. Have students do exercises 3.1.2a, 3.1.5a, and 3.1.5b.

5 Use the computer program *EM Field* to illustrate the pattern of electric field near single positive and negative charges (see problem HW3-7).

**Week 3**

L1 5 Brief review of definition of electric field.

40 Discuss superposition. Work out initial calculation in section 3.2.2. Emphasize drawing the individual field vectors and calculating *positive* magnitudes of these vectors. Then resolve into components. The key point is to avoid sign errors by dealing almost solely with positive quantities and using the drawing to get the component signs. Students do 3.2.2a, 3.2.2b, followed by discussion. Then work out electric field of dipole.

5 Use the computer program *EM Field* to illustrate the pattern of electric field near a dipole (see problem HW3-7).

R1 50 Discussion of Chapter 2 homework assignment (students turn in homework at start of class). We feel that problem HW2-1 is the most significant problem, because it involves so much good physics and chemistry. There are two key issues: spark formation through a chain reaction involving free electrons in the air, and polarization of the body (and the doorknob) which are crucial to producing electric forces in the air between finger and doorknob sufficient to ignite the chain reaction. Note that the polarization of the body leading to excess charge on the finger is probably due not so much to molecular polarization but rather to movements of the sodium and chloride ions in the blood and in the salt water on the skin.

In the discussion lead the students to see that as the finger approaches the doorknob the force on a free electron in the air increases for *two* reasons: the nearby charges are closer to the electron, and these nearby charges are larger due to the mutual polarization of the finger and the doorknob. When the force gets large enough to trigger the chain reaction, you get a spark.

It is great fun to think through the species involved in charge transfer between finger and doorknob. For example, if the sole of the shoe is charged negatively, polarization will make the finger have a negative charge (higher concentration of chloride ions), and the facing portion of the doorknob will be positive. When the spark forms and the air becomes conducting, electrons fall onto the doorknob, and positive ions migrate to the finger. A possible reaction is for a nitrogen ion to take an electron from a chloride ion and liberate chlorine! Of if the finger is positive, an electron attracted to the finger can neutralize a sodium ion which then reacts with water to form sodium hydroxide and liberate hydrogen! Students should be encouraged to use U and L tapes to test the sign of the charge on the bottom of the shoe.

The net effect of the charge transfer in the spark is to reduce the charge on the near side of the doorknob and on the finger, eventually reaching a level below that necessary for the chain reaction, and the spark goes out.

L2 50 Force due to permanent dipole and due to induced dipole; have students work out some of the exercises before discussion. Students carry out the experiment of observing attraction when two like-charged tapes get very close to each other. Most teams did not see the desired effect, but they had seen it twice before: once when charging a metal foil by contact with a charged pen, and once with the Van de Graaff generator.

It became clear during the lecture that many students were rather unsure of the field concept, and it was not possible to go as far during the lecture as had been planned. This was in part due to major disruptions in the previous week due to severe winter weather that shut down the university. Because the field concept is the backbone of the entire course, we decided to slow down and make sure that the material of Chapters 3 and 4 is well learned. We'll skimp on other less central topics later.

R2 20 Students work in small groups on review questions RQ3-10, 3-1, 3-2, and 3-8, and on exercise 3.3.4a.

15 General discussion of these questions.

15 Students work in small groups on one of the assigned homework problems due next Tuesday.

L3 15 Short quiz on very basic concepts in Chapter 3.

35 Final summary of Chapter 3 including limitations of the field concept, and the reality of the field due to retardation phenomena. For the latter, had a student pretend to hold up a charge, and I held up a pencil pointing away from that charge, about ten feet away. Then the student moved the charge sideways, and I counted out loud "1, 2, 3, .....10" at which point I shifted the pencil to point away from the new position of the charge. During those 10 "nanoseconds" my measurements of the electric field continued to point away from the old position of the charge. Discussed (unassigned) problems HW3-2 and HW3-3.

**Week 4**

L1 50 Introduce the topic of calculating the field due to distributed charges. Went through the four-step procedure for a uniformly charged rod. Began with asking students why we do this, since few people are in fact particularly interested in the electric field of a rod, and fortunately a student volunteered that it must be that this is practice for more relevant cases. Was able to get all the way through the four-step procedure, including getting the students to invent tests to check the integral-calculus results.

R1 50 Discussion of Chapter 3 homework assignment (students turn in homework at start of class). Problem HW3-8 is particularly interesting because a trivial experiment gives us information about atomic polarizability (it is useful to do the experiment: rub a plastic pen through your hair, then pick up a small scrap of paper). The problem also provides a useful exercise in keeping things straight. For example, it is useful to ask not only about the field due to the dipole at the location of the pen, but also about the (very much larger) electric field due to the pen at the location of the dipole, which nevertheless yields exactly the same force (Newton's third law). In problem HW3-6, emphasize the relative magnitudes of the various electric field contributions: the field at the center is much smaller than the field at the left and right ends.

L2 50 Quickly summarize a ring, go through disk in detail. Discuss capacitor (two disks); students do 4.5a and 4.5b, which are at the heart of the matter.

R2 10 Students work in small groups on review questions RQ4-2, RQ4-6, RQ4-8; don't expend too much time on this.

15 General discussion of these questions.

10 Students work in small groups on one of the assigned homework problems due next Tuesday. Appropriate problems are HW4-4 and HW4-5.

15 Discuss the problem together.

L3 20 Quiz on finding the field due to a charge distribution (a ring is an appropriate one, or else give the necessary integral formulas).

30 Spherical shells of charge. Applications, with emphasis on problems which involve two basic principles: superposition, and zero field inside a metal in static equilibrium. Again demonstrate Van de Graaff generator; charges can easily migrate from the belt to the outer metal shell, because the field inside the shell is small; refer to problem 4-8 for breakdown criterion. Ask students to do problem HW4-10 (direct ionization).

**Week 5**

L1 10 Summarize for the first hour exam, but instead of listing all the topics of the first four chapters, structure the summary in terms of lists under two basic headings:

| Charges make electric field | Electric field affects charges |
|---|---|
| charge conservation | $\vec{F}_2 = q_2\vec{E}_1$ |
| $\vec{E}$ of various charge distributions: | superposition principle |
| point | interaction through matter |
| $\vec{E} = \frac{1}{4\pi\varepsilon_0}\frac{q}{r^2}\hat{r}$ | interaction not instantaneous |
| dipole | various effects of $\vec{E}$: |
| far away, approximately $1/d^3$ | attraction and repulsion |
| induced dipole force on point charge $1/d^5$ | polarization |
| uniform rod | insulators ($p = qs = \alpha E$) |
| near, approximately $1/d$ | ionic solutions |
| uniform ring | metals |
| uniform disk | $E = 0$ inside metal in static equilibrium |
| near, approximately uniform | chain reaction in air ($E_{critical}$ about $3\times10^6$ N/C) |
| capacitor (2 uniform disks, +Q and -Q) | |
| uniform spherical shell | |
| from outside, like point charge | |
| inside, shell contributes zero field | |

40 Work problem HW3-8. This is a very comprehensive problem, touching on many of the topics of the first four chapters. If there is extra time, do problem HW4-8.

R1 50 Discussion of Chapter 4 homework assignment. Because there is an exam on the following day, we loaned red pens and had the students correct their own work. Credit was given for attending.

L2 Evening exam, and no class for L2, but we were present in the classroom to answer questions. Nominally an "hour" exam, but students were given at least an hour and half to eliminate time pressure. The exam is included in this instructor manual.

R2 50 Pass out electricity kits (without magnets at this time), have students start on Chapter 5. Tell students to bring their kits to class tomorrow.

It is extremely useful to record on the blackboard the results of each group as they obtain data. These public data provide an excellent basis for general class discussion as to why results of different groups are the same, or are different.

L3 45 Students continue working on Chapter 5, in groups of two. Again, record on the blackboard the results obtained by the groups, and break in when appropriate for brief general discussion to compare results.

5 Useful to show burn-up of cracked round bulb. Point out that bulbs typically contain argon rather than vacuum—one student suggests a compelling reason: probably easier to manufacture. Perhaps the argon also inhibits vaporization of the tungsten onto the glass? Point out the importance of the pattern of magnetic field around a wire, and that we have never seen an electric field pattern that was curly like that.

Hand back exams, with full solutions and indications of scoring for partial credit. No class time was spent in going over the exam.

**Week 6**

L1 25 Short quiz on Chapter 5, or have students submit their summary results from page 191 and page 201. Review the key puzzles of Chapter 5, then discuss the important formulas $i = nAv$, $v = uE$, and $i = nAuE$. These formulas should be memorized, though it is even better to be able to recapitulate the derivation of $i = nAv$. Discuss the free-electron model for metals, and the stop-start motion that results in an average drift speed.

At this time we are omitting section 5.6 on additional aspects of magnetic field. We may come back to this later, when we study magnetic fields. However, it may be very useful at least to touch briefly on the superposition principle for magnetic field illustrated at the start of section 5.6.

25 Begin Chapter 6, discussing through section 6.2.1, which reveals where the charges are that make the observed steady-state field.

R1 50 Continue discussion of Chapter 6, through the treatment of the thin resistor (section 6.5.1).

L2 50 Review the thin resistor, then discuss the wide resistor and parallel resistors (sections 6.5.2 and 6.5.3). Then discuss the mechanical battery and the energy argument, leading to the central conclusion that $E$ is determined by the work per unit charge done by the battery, and the current is given by $i = nAuE$.

R2 30 Applications of the theory, involving experiments. Students work in groups of 2 or 3 on sections 6.7.1 (length), 6.7.2 (area), and 6.7.6 (two batteries vs. one). We put the current conservation and energy conservation equations on the board and encouraged the students to work out the theory before making the measurements. As students got numerical values these were recorded on the board for class comparison. We went around and quizzed groups about the theoretical predictions, and we ended with general discussion of the work, insisting on rigor in reasoning from the energy and current equations to predict the observed results.

Note: students tend not to read directions very carefully; the length and area measurements should be done as stated (long wires to make sure that the wire resistance is large compared with the internal resistance of the batteries, thin wire in the length

177 and 195?

measurement for similar reasons, one battery to keep the compass deflection below 20 degrees).

10 Students (still in groups) discuss selected review questions (RQ6-1, 6-2, 6-6, 6-10).

10 General class discussion of the review questions.

This was a very good and constructive class hour that helped solidify the key points in Chapter 6.

L3 15 Short quiz on very basic concepts in Chapter 6.

35 Work through some Chapter 6 homework problems.

**Week 7**

L1 40 Students start working together on Chapter 7 (capacitor circuits). We loaned students a half-farad capacitor to be used during their study of Chapter 7, and then returned to us.

We find it useful to have the students do these experiments together in class. Not only does this insure that they actually carry out the critical experiments. It also provides a social context within which to argue with each other in small groups about their observations and analyses.

10 Short discussion of what they have just seen: a non-steady-state current, leading to a final state of static equilibrium.

R1 50 Discussion of Chapter 6 homework assignment (students turn in homework at start of class).

L2 45 Finish the experiments in Chapter 7, leaving theoretical aspects to be completed outside class.

5 Point out that the non-steady-state currents they observe represent a very slow approach to static equilibrium. Note that breaking a circuit (or one branch of a parallel circuit) is a similar process: current continues to go down the wire and piles up at the gap, eventually stopping the flow. So a gap in a circuit is just a very small capacitor.

R2 50 Discussion of short Chapter 7 homework assignment (students turn in homework at start of class). Part of the assignment included collecting some work from Chapter 7 (sections were not announced before class). We had students turn in summaries of their work in section 7.4 (the effect of different bulbs on charging) and their work in section 7.6.3 (role of the insulating layer).

The class discussion of the problems was excellent, but later that day we saw that their written explanations of the processes in RC circuits were weak and inadequate, so we planned something special for the next class (L3).

L3 10 We displayed examples of poor and good student explanations from the homework collected on the previous day. We stressed the need for attention to detail and good diagrams, and emphasized that we would be looking for improvement on the quiz.

15 Short quiz on basic concepts in Chapter 7. The results of this quiz were quite good. It seemed that the combination of having worked on the homework problems, having discussed them in class, and seen a comparison of good and poor solutions, resulted in good performance on the quiz.

25 Began discussion of Chapter 8 (electric potential). Review basic mechanics principles of work, and of potential energy which can be measured as equal to the external work if the kinetic energy of the system doesn't change. Then discuss quasi-static processes, first with respect to lifting a weight, and then with respect to moving a charge across a capacitor while exerting an external force to prevent change in kinetic energy. Finally define electric potential as the potential energy change per unit charge.

**Week 8**

L1 no class—midsemester holiday.

R1 50 This recitation class was run rather like the lectures. We continue working through Chapter 8, hitting the high spots; students are expected to go back over the material and fill in the details. While all of this is in the book, our experience is that the concept of potential is quite difficult for students, and so it helps to see and hear this material several different ways. Stress the importance of doing all the in-line exercises to get these difficult concepts solid.

Specifically, we quickly reviewed the basic concept and the simple rule for the sign of $\Delta V$ (decreasing if going in direction of field). Then we showed that for a "short enough" displacement (one for which the field is nearly uniform), $\Delta V = -E_{parallel}\Delta l$. We then skimmed the four-step procedure for calculating $\Delta V$, noting the similarities to the four-step procedure for calculating electric field from distributed charges, but with the difference that there is a path involved (because $\Delta V$ involves two locations, whereas field is calculated at a single location). Rather than dwelling very long on the general procedure, we immediately applied it to the concrete cases of the circuit and the point charge (sections 8.3.2 and 8.3.3).

We stated that tomorrow we would see that we would get the same result no matter what path we took, so we would choose a path that simplifies the calculation. In the case of the circuit, we asked why we wouldn't want to take a path that was partly in the wire and partly in the air, getting the students to say that this is in principle a possible path but one along which it would be difficult to evaluate $\Delta V$ because of ignorance about the field in the air.

L2 50 Path independence: Students did parts of exercise 8.4a, with general discussion. We skimmed the general derivation and concentrated on the exercises 8.4.1a and 8.4.1b, and the paired results of path independence and the fact that the round-trip $\Delta V = 0$. Students worked for a few minutes on exercise 8.4.2a (the curly electric field), followed by discussion. One group suggested that a simple circuit is an example of a curly electric field pattern, which provided the opening for showing such a circuit and seeing that inside the "mechanical" battery the electric field is large and points in the other direction, so that the round-trip $\Delta V$ is still zero. Another group mentioned the field that curls around a current-carrying wire but were reminded that it was a *magnetic* field. Finally another group gave the argument that the curly electric field has a non-zero round-trip $\Delta V$ and so cannot be produced by point charges. We see that the potential concept lets us rule out certain patterns of electric field, which is a strong constraint. All this is preamble for the surprising curly non-Coulomb electric field to be encountered in Chapter 13 on magnetic induction.

Summarized the argument for $\Delta V = 0$ throughout a metal in static equilibrium. Stressed that it is very important that students work through all the in-line exercises in this chapter, including those in section 8.5.

Finally showed that electric field is the gradient of the potential (section 8.6).

R2 50 Discussion of Chapter 8 homework assignment (students turn in homework at start of class), involving some of the easy problems (8-1 through 8-6). These homework problems are easier than the usual homework problems and are done at this time to nail down the most fundamental aspects of electric potential.

L3 15 Short quiz on basic concepts in Chapter 8.

35 Continue Chapter 8—potential difference across multiple regions (section 8.8). In our experience this causes students the most trouble of all the topics in the chapter. It calls for careful use of the standard four-step procedure, including a good diagram of the pattern of electric field.

**Week 9**

L1 25 Finish Chapter 8 (potential of a point charge relative to infinity). After a lot of experimentation, we have concluded that it is dangerous in the introductory course to dwell on potential as q/r because it is enormously confusing to beginning students and undercuts the important principle that it is potential *difference* that matters. For that reason throughout section 8.9 we continually write $V - V_\infty$ rather than V.

5 Briefly use *EM Field* to visualize potential relative to infinity and potential differences near a point charge (may want to turn on the Constrain to Grid option and then pile a bunch of large charges on top of each other to get a large field near the charge).

20 Go over selected Chapter 8 review questions, especially the first one, which has some traps that many students will fall into. Point out that it is useful to fall into these traps once in order to focus on some subtle distinctions that have to be made among the various quantities. As usual, have students work in small groups for a few minutes before discussing the review questions.

R1 50 Discuss Chapter 8 homework problems (students turn in homework at start of class). It may be useful to point out that a rather common mistake is to evaluate an integral of a non-uniform field as follows: "$\Delta V = -\int E dl$ can't be written as $-E\int dl$ because E is varying, so it must instead be $-(E_f - E_i)L$." Of course this does violence to the notion of integration, and the mistake can be avoided by focusing on the fact that it is not just the values of the electric field at the ends of the interval that matter, but all values of E along the path. This point is also covered in RQ8-1e.

L2 30 Summary of the transition from electrons/sec = nAv = nAuE to conventional current I (coulombs/sec) = qnAuE, J (amp/m$^2$) = (qnu)E = $\sigma$E, and $I = |\Delta V|/R$, with $R = L/(\sigma A)$. This is pretty familiar to the many students who studied circuits in high school, but one must be careful not to leave the other students behind, and both groups need to see how the microscopic view of circuits is related to the macroscopic view.

Many students get an overdose of Ohm's "law" in high school (typically as V = IR rather than $I = |\Delta V|/R$), and they confuse this empirical statement about materials (at constant temperature!) with the much more fundamental energy-conservation equation (the Kirchhoff loop rule). As a result, confusion abounds. For example, when the students were doing experiments in Chapter 5, one student said that in a circuit containing a battery and a bulb, the current was the same in the wires going into and coming out of the bulb, but the current in the bulb filament had to be much less, because V = IR, and in the bulb filament the R was much bigger but the V was the same.

One thing we do to avoid some confusion is that we don't use the term "Ohm's law" at all; rather we simply refer to "ohmic" and "non-ohmic" materials. We emphasize that light bulbs and batteries aren't ohmic, in the students' own experience.

20 Discussed the details of emf as non-Coulomb work per unit charge ($F_{NC}d/e$ for the mechanical battery), and the fact that the sum of the (Coulomb) potential differences around the circuit is zero, because the electric field inside the battery points in a direction opposite to the field in the wires.

R2 50 Loop and node equations. Power. Internal resistance—students should bring experiment kit to be able to do the experiments on internal resistance (section 9.4.2). It is appropriate to discuss review questions RQ9-6 and RQ9-9.

L3 50 Multi-loop circuits. Ammeters and voltmeters—students should bring experiment kit to be able to do the experiments on ammeters in section 9.5.

Students sign up for a two-hour walk-in lab in which they bring their own experiment kits to which we add a pair of digital multimeters (Friday and Monday). The goal of this lab is to gain experience in the use of digital multimeters and to measure some properties of circuits much more accurately than is possible with a compass. The lab report form is reproduced later in this instructor manual.

**Week 10**

L1 20 Demonstration of surface charge on a high-voltage circuit: four 22-megohm resistors in series, using two 5000-volt floating power supplies in series to make one end of the resistor chain be at +5000 volts and the other end be at -5000 volts relative to ground:

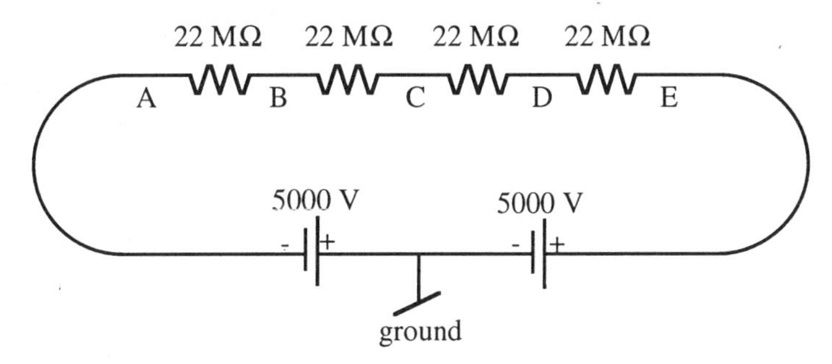


The resistor chain is supported in mid-air by punching the ends of the wires through the tops of two Styrofoam blocks. This gets away from the table and possible unwanted effects from polarization of the table. First discuss the circuit, paying attention to surface charge, field, and potential. Then bring a pith ball coated with conducting paint near the negative end of the resistor chain (point A) and see it first attracted, then repelled (after being charged by contact). Rub plastic pen through hair and show that it repels the pith ball, proving that the pith ball did indeed get charged negatively. Repeat with positive end of resistor chain (point E): positive charge on pith ball. In the middle of the chain (point C) nothing happens (no surface charge there). At point B or D we get a medium effect. The point of all this is that we need a very high voltage to get enough surface charge to be able to see mechanical effects of that charge.

30 Discuss RC charging circuit from point of view of qualitative graphical integration, based on the energy-conservation loop rule. Make graphs of Q(t) and I(t). This complements the numerical and calculus treatments in the book.

R1 50 Discuss Chapter 9 homework problems (students turn in homework at start of class).

L2 Evening exam, and no class for L2, but we were present in the classroom to answer questions. Nominally an "hour" exam, but students were given at least an hour and half to eliminate time pressure. The exam is included in this instructor manual.

R2 50 Start Gauss's law. Parallel the book, placing emphasis on the qualitative patterns of electric field, and on developing the definition of electric flux. It is then possible in this one period to give a once-over-lightly proof of Gauss's law.

L3 40 Quickly review Gauss's law, then apply it to prove the two assertions that we had made without proof early in the course: 1) the field of a uniform spherical shell of charge (section 10.5.2), and 2) the fact that there is no excess charge in the interior of a conductor (section 10.5.4). Next, find the electric field of a long uniformly charged rod, and show that we get the same result very easily that cost us a lot of effort in Chapter 4.

10 Used the *EM Field* program to visualize Gauss's law applied to irregular cylinders drawn around long charged rods. It is useful to show many different situations, including showing how a charge outside the Gaussian surface does contribute to the flux all over the surface but makes a *net* contribution that is zero. Note that the program lets you move

charges, or add additional charges, and then immediately replots the new flux on the old Gaussian surface, which makes it very easy to illustrate the role of the outside charges, which without this visualization often causes lots of conceptual difficulties for students.

In some semesters we spend more time on Gauss's law. This semester some topics had to be squeezed a bit due to upheavals associated with bad winter weather.

Hand back exams, with full solutions and indications of scoring for partial credit. No class time was spent in going over the exam.

**Week 11**

L1 50 Students started working on Chapter 11, doing desktop experiments with partners. We recorded on the blackboard student results for the distance at which the compass deflection was 60 degrees, the deflection at twice this distance, the ratio of the magnetic fields at these two distances, and the exponent n if the field falls off like $1/d^n$. Although the distance for a 60-degree deflection varies quite a bit (indicating that the magnets aren't all of the same strength), most groups do get a value of n around 3.

The emphasis this week in class is on the experiments on magnetism in Chapter 11, with much of the experimental work done in class. Perhaps we did not put quite enough stress on the theory in class, but it is very important that students have direct experience with the phenomena and issues in simple situations.

R1 25 Go over assigned Chapter 10 review questions. Extend the question about the flux on the box (RQ10-3) to ask students to design a charge distribution that could produce such a pattern of electric field. It isn't easy, but they do eventually come up with a positive sheet slicing through the box, and another positive sheet to the left of the box. One can see that this isn't unique: there could instead be a negative sheet to the right of the box.

15 Students work together in groups of two or three on homework problem HW10-2.

10 General discussion of problem HW10-2. One of the more interesting points to make is that the fringe field that appears in Gauss's law for Gaussian surface 4 is obviously due to all four sheets of charge, yet the charge that appears in Gauss's law is just a piece of the charge on the left-most face. This is a particularly nice example of the strange relationship between field due to all charges and net flux due only to charges inside the surface, a mystery well explained by the program *EM Field*.

L2 50 Brief description of the nature of the earth's magnetic field at our location (about a 70-degree dip angle in Pittsburgh, pointing into the earth), and the fact that the horizontal component of the magnetic field, which affects the compass, varies quite a bit over the earth's surface. Show a dip compass.

Brief review of the vector cross product, using the example of the field near a current-carrying wire, and showing that the direction given by the cross product is the same as given by our old right-hand rule. Asked what version of the right-hand rule the students were familiar with. Emphasized that we will be doing a lot with the right-hand rule, and that the version presented in the book has the very big advantage that it captures the angle whose sine comes into the magnitude of the cross product, whereas right-hand rules that emphasize three perpendicular axes don't provide this important information.

Magnetic field of a long straight wire, followed by use of the computer *EM Field* program to show the magnetic field near a wire.

Students do the experiment on a long wire (section 11.3.1), with results tabulated on the blackboard.

If there is time, quick overview of the magnetic field of a current loop.

R2 50 Students work in groups of two or three on measuring the magnetic field near a current loop (section 11.4.3). Results from groups are tabulated on the blackboard. They also do section 11.5; it is important that they record the magnetic moment of their own magnet, which will be used for various purposes later on. (During this class period, skip sections 11.4.4, 11.5.1, and 11.5.2, which students should however work through outside class.)

L3 15 Short quiz on basic concepts in Chapter 11.

35 Work through section 11.6 and compare theory and experiment.

Discuss ferromagnetism (sections 11.6.1-11.6.2).

Hand out unmagnetized nails (2 or 3 per student) so that students can do experiments on ferromagnetism outside class (exercises 11.6.1a-d).

Between week 11 and week 12 there was one week of spring vacation.

**Week 12**

L1 30 Using a see-through CRT (which lets you see the curving trajectory along a phosphor plate) or an oscilloscope, demonstrate magnetic forces on an electron beam, using a bar magnet and a current-carrying coil. Get the students to "invent" $\vec{F} = q\vec{v} \times \vec{B}$ from the observations. (One does have to say that there is no coefficient k because the unit of magnetic field, the tesla, has been adjusted appropriately.) Discuss the full Lorentz force. Discuss the basics of circular motion. Emphasized the importance of doing the Lorentz-force exercises in the book to be very familiar with it.

20 Discussion of the Hall effect. This discussion draws together many strands of the course. There is a big payoff from previous work. In particular, students are very familiar with drift speed of individual electrons, and with the concept of a dynamic equilibrium (steady state) that develops with charge buildup on the sides of the conductor. It is useful to ask the students why there is no transverse electric field in the absence of a magnetic field, despite the presence of surface charges.

R1 50 Discuss Chapter 11 homework problems (students turn in homework at start of class).

L2 25 Magnetic forces in moving reference frames (section 12.3). Students really enjoy this discussion, with its remarkable climax concerning differences in time intervals. This is a high point of the course for teachers and students.

25 Demonstrate the magnetic forces on current-carrying wires: a wire lying in the gap of a large magnet jumps when the current is turned on; the wire curves through the magnet gap, supported by the field; a single loop of this wire twists in the magnet gap; a coil rotates on an axle into the stable position; a coil with a commutator rotates continuously on an axle, acting as a motor (rotates more slowly when moved away from the magnet—ask why, and also ask where energy to run the motor comes from); two parallel wires with current flowing parallel or anti-parallel attract or repel each other. Mention that there is a connection between the twist of a coil and the twist of a compass needle. All of these demonstrations are analyzed merely qualitatively in terms of the Lorentz force on individual electrons (or conventional positive carriers) in the wire. In the case of the parallel wires, it is important to emphasize the two-step nature of the reasoning: one wire makes a field, and the other wire is affected by that field.

We did not do the derivation of $d\vec{F} = I d\vec{l} \times \vec{B}$ from $\vec{F} = q\vec{v} \times \vec{B}$ in class but pointed to the derivation in the book and commented that this is exactly similar to what we did with the Biot-Savart law.

R2 25 Students work in small groups on selected review questions (RQ12-1, 12-2, 12-3, 12-4, 12-5).

15 General discussion of these questions.

10 Students work in small groups on an assigned homework problems (HW12-10) due next Tuesday.

L3 15 Short quiz on basic concepts in Chapter 12.

30 Discuss twist on a current-carrying loop, potential energy of a magnetic moment, and the force on a magnetic moment in terms of change of potential energy in a non-uniform field. Left some related topics to be studied in the book.

5 Provide time for students to measure distance at which one magnet will pick up another (exercise 12.5.3e). They also need to record the magnetic moments of both magnets (which were measured in exercise 11.5a) in exercise 12.5.3c. They should analyze the data outside class.

**Week 13**

L1 50 Begin Chapter 13, magnetic induction. Start by summarizing the properties of a long solenoid. Use a qualitative Biot-Savart argument for the magnetic field being small outside (far from the ends), then demonstrate this by placing a transparent compass on the overhead projector and bringing the end of the solenoid near the compass, then bringing the side of the solenoid near the compass. Ask about electric field and make sure that students understand that except for the very small electric field associated with the very small amount of surface charge, the space around a DC solenoid is nearly free of both electric and magnetic field.

Next state as an experimental observational fact that outside a long solenoid there is a curly non-Coulomb electric field when B inside the solenoid *changes*. Ask what is odd about this curly electric field, and wait for a student to point out that the round-trip path integral of this electric field is nonzero. Ask what would happen if we put a copper ring around the solenoid—would drive a current. Then do the demonstration. We had a transparent galvanometer on the overhead projector connected to a secondary coil surrounding the midsection of a long solenoid. The solenoid was connected to a variable power supply, and we experimented with turning the current up and down. The emphasis here is on the fact that nothing happens with a large steady current; the effect is proportional to dB/dt (slow and fast changes make small and large currents run in the secondary).

Went through the calculation of emf for a circular path around the solenoid, and for a circular path of twice the radius (same because $E_{NC}$ is goes like 1/r). Reviewed the calculation in the book for emf when the secondary does not surround the solenoid, then demonstrated that indeed little happened (students correctly explained the small remaining effect as due to the small fringe field of the solenoid).

Removed the solenoid, and experimented with moving a bar magnet near the secondary. Also drove a primary coil along the same axis as the secondary coil. Demonstrated what happened at other angles, especially at right angles (with either the bar magnet or the primary coil), in which case it is nice to see that $E_{NC}$ is nearly transverse to the wires and so drives little current. Also looked at the effect of moving the two coils farther apart and changing the primary current.

Discussed and used the right-hand rule for induction (thumb in direction of -dB/dt, fingers curl around in the direction of $E_{NC}$).

A student asked, "*Why* does dB/dt give an electric field?" This was a perfect straight line to discuss the nature of physical laws, as discussed in the book.

R1 50 Discuss Chapter 12 homework problems (students turn in homework at start of class). This should include the results of exercises 12.5.3c-12.5.3e dealing with the prediction and measurement of one magnet picking up another.

L2 50 Discussed and practiced with Lenz's rule, in the context of the same equipment that we had used in the preceding lecture. Went through section 13.3.3 on a quantitative calculation of the effect of one coil on another.

Went back to the demo of the long solenoid with a secondary coil around its midsection. With an iron bar placed inside the solenoid, asked the students how the presence of the iron would alter the effect. Alas—no one could see that the key point would be a larger magnetic field (amplification effect of aligning the magnetic domains in the iron), and hence a much larger effect. (Probably no one had done exercise 11.6.1a, which was to be done outside of class, but with intervening spring vacation.) Then we ran the primary current up and down and showed that there was indeed a much larger effect in the secondary.

R2 20 Discuss selected Chapter 13 review questions and homework problems Choose problems that emphasize actually calculating an emf, not just qualitative analysis. A specific difficulty we see is that some students incorrectly view Lenz's rule as a *quantitative* law—that the amount of current that runs is exactly the amount necessary to cancel the flux change, something that is true only for superconducting circuits.

L3 no lecture (Spring Carnival)

Students sign up for a two-hour walk-in lab on magnetic induction (Mon. and Tues. of week 14).

### Week 14

Monday and Tuesday afternoon: sign-up lab for students to come in for a two-hour magnetic induction lab. Using an oscilloscope as a sensitive voltmeter, students observe the emf developed in a large coil under various conditions: moving their little bar magnet toward or away from the coil; changing the current in a neighboring coil (using an AC power supply with variable frequency); observing the emf across a coil in an AC RL circuit; observing an LC circuit, both free oscillations (triggered repeatedly by a square-wave generator) and resonance in a driven circuit. Students bring their own kits; the experiment uses a battery and the bar magnet (and the value of its magnetic moment, measured previously in exercise 11.5a). The lab report form is reproduced later in this instructor manual.

L1 50 Discuss motional emf: moving bar alone; moving bar in circuit; moving loop; rotating loop. Demonstrate a homemade coil rotating in a large magnetic field to light a bulb. Use a commercial hand-held generator to light a bulb. Have a student crank as fast as possible; make and break the connection to give a kinesthetic feel for the back force. Pass the generator around the class to let everyone have the experience.

Superconductor discussion and demonstration of zero resistance and the Meissner effect. The Meissner effect is illustrated by floating a magnet above a superconducting disk. Classically, one cannot float an unconstrained magnet in a stable position. Students may have seen a science toy consisting of a rod that floats horizontally (and can be spun on its axis) but must have one end against a vertical surface to be stable. Without the Meissner effect, a quantum-mechanical effect, the demo of the unconstrained floating magnet would be impossible.

R1 50 Discuss Chapter 13 homework problems (students turn in homework at start of class).

L2 50 Eddy currents demonstration: A solid and a slotted copper pendulum swing through a magnetic field. The slotted pendulum has narrow eddy currents and experiences a correspondingly small net magnetic force on the two branches of the eddy.

RL circuit: The book discusses RL circuits in formal math terms (solve the differential equation). To supplement this, we analyzed an RL circuit by writing the energy-conservation loop equation and solving for $dI/dt = (emf - RI)/L$, then using this instantaneous value of $dI/dt$ to analyze the evolution of the system graphically. The initial current is 0 (can't change instantaneously, because that would mean an infinite voltage across the inductor), but the initial dI/dt = emf/L, which leads after a short time $\Delta t$ to some I = (emf/L)$\Delta t$. Now that there is some I, dI/dt = (emf - RI)/L is smaller, so I changes less in the next time interval. By graphing I and dI/dt vs. t, one can see the exponentials evolve.

Had a demo of a BIG inductor connected through a knife switch to a storage battery (low internal resistance), and showed the big spark that occurs when one tries to break the circuit.

Had intended to discuss LC circuits but ran out of time; directed the students' attention to the discussion in the book. Note the LC circuit work in the magnetic induction lab.

R2 50 Work some Chapter 13 review questions (RQ13-6, 13-7) and homework problems (HW13-5, 13-9, and if time 13-4).

L3 50 Review for exam.

**Week 15**

L1 Evening exam, and no class for L1, but we were present in the classroom to answer questions. Nominally an "hour" exam, but students were given at least an hour and half to eliminate time pressure. The exam is included in this instructor manual.

Usually we devote a full week to Chapter 14, but snow days and other abnormal conditions led to having less than a full week available in Spring 1994. As a result, in recitation section R1 we skimmed quickly over Ampere's law and its generalization by Maxwell, and we briefly summarized the discussion of the moving slab of radiation. All of this is of high cultural interest, but we are most interested in getting as quickly as possible to what we consider the most important aspect of the chapter—the production of electromagnetic radiation by accelerated charges, and the interaction of electromagnetic radiation with matter. After this compressed R1 class, L2 was as summarized later.

R1, compressed version (combination of normal L1 plus R1):

30 State that we're going very quickly over sections 14.1-14.3.4—see details if interested.

Skim Ampere's law—analogy with Gauss's law; show "piercing" diagrams (section 14.1.2).

Apply Ampere's law to thick wire (exercise 14.1.3a); mention other applications.

Extending Ampere's law: diagram in section 14.1.4; do exercise 14.1.4a with students.

Maxwell's equations (section 14.2).

15 Summarize moving slab (sections 14.3-14.3.5); emphasize final result: E, B, and v (=c).

Quickly dispose of electric and magnetic versions of Gauss's laws.

Do Faraday's law in some detail, getting E = Bv.

State that Ampere-Maxwell law is similar, leading to $B = \mu_0\varepsilon_0 Ev$.

Punch line—solve for v:

$$v^2 = \frac{1}{\mu_0\varepsilon_0} = \frac{1}{\left(\frac{\mu_0}{4\pi}\right)(4\pi\varepsilon_0)} = \frac{9\times10^9}{10^{-7}}\frac{\text{m}^2}{\text{s}^2} = 9\times10^{16}\frac{\text{m}^2}{\text{s}^2}$$

Emphasize that *only* this configuration of E, B, and v satisfies all the Maxwell equations.

5 Effect on matter—sections 14.4-14.4.4.

Next we reproduce what we do when we have a full week:

L1 50 Sections 14.1-14.2: Ampere's law, and its generalization by Maxwell to include the displacement current. Use the computer program *EM Field* to illustrate Ampere's law. Applications of Ampere's law. Maxwell's extension to Ampere's law. Maxwell's equations.

R1 50 Sections 14.3-14.3.5: Consider a slab of space (thickness d and infinite extent) within which there are uniform crossed E and B fields, and assumed to be advancing at a speed v. Using rectangular paths for Faraday's law and the Ampere-Maxwell law that are partly in and partly out of the slab, show that this configuration of fields is compatible with Maxwell's equations, if the speed is c.

Sections 14.4-14.4.4: Discuss the effects that such a slab would have on matter: main effect is due to E, but since E gives a charge some transverse velocity there is a forward magnetic force (radiation pressure).

L2 50 How can we produce this field configuration? Use a computer program to illustrate how an accelerated charge produces a pattern of electric field with a transverse kink: the key diagrams are shown in section 14.5 of the book. Discuss the polarization character of this field at different orientations to the acceleration, showing that the propagating transverse electric field is proportional to the perpendicular component of the acceleration. Demonstrate the effects with a radio transmitter and antenna that lights a flashlight bulb (like that shown in section 14.7.3 of the book). Also demonstrate a crude polarizer consisting of a cardboard sheet with parallel wires, which when placed between transmitting and receiving antennas illustrates the role of polarizers. Also demonstrate polarization of visible light.

Hand back exams, with full solutions and indications of scoring for partial credit. No class time was spent in going over the exam.

R2 15 Students work in small groups on selected Chapter 14 review questions (RQ14-3, 14-4, 14-5).

10 General discussion of these questions.

15 Students work in small groups on additional review questions and homework problems (RQ14-6, RQ14-9, HW14-5).

10 General discussion of these questions.

L3 15 Short quiz on basic concepts in Chapter 14.

25 Why the sky is blue, and polarized (section 14.7.4).

10 Students fill out final course evaluations.

Final exam week: Three-hour final exam covering the entire course. Students were given as much as four hours if needed, to eliminate time pressure. They were encouraged to bring one page of notes, with the main purpose to help them organize their thoughts while reviewing. We also provided a formula sheet. The exam is included in this instructor manual.

# CHAPTER 6

# SAMPLE QUIZZES & EXAMS (SEMESTER A)

Here we present quizzes, exams, and the final exam for the same semester documented in detail in previous chapters (semester "A"). The weekly quizzes preceded the homework assignment for a topic and were quite easy, testing only the most fundamental aspects of the topic.

Quizzes were nominally 15 to 20 minutes. Exams were nominally an hour long, but those students who needed more time were given an additional half-hour or hour. The final exam was nominally three hours long, but those students who needed more time were given an additional hour or two.


Quiz for Chapter 1 68

Quiz for Chapter 2 68

Quiz for Chapter 3 68

Quiz for Chapter 4 69

Quiz for Chapter 6 69

Quiz for Chapter 7 70

Quiz for Chapter 8 70

Quiz for Chapter 11 71

Quiz for Chapter 12 71

Quiz for Chapter 14 72

Exam 1 covering Chapters 1 through 4 73

Exam 2 covering Chapters 5 through 9 77

Exam 3 covering Chapters 10 through 13 81

Final exam covering Chapters 1 through 14 85

## Quiz for Chapter 1

**Question 1:** When you rub a plastic comb on wool, the comb becomes negatively charged. For each of the following, state whether each could possibly be the dominant charging mechanism or not, and briefly explain why.

a) Individual protons are transferred from the nuclei of some of the atoms in the comb to the wool.

Yes/no:__________ Briefly explain why:

b) Individual electrons are transferred from some of the atoms in the wool to the comb.

Yes/no:__________ Briefly explain why:

c) Positive ions are transferred from the comb to the wool.

Yes/no:__________ Briefly explain why:

**Question 2:** When two protons are a distance x apart (where x >> proton size), they each exert a force $F_1$ on each other. When a carbon nucleus (6 protons) and a beryllium nucleus (4 protons) are a distance 3x apart, they each exert a force $F_2$ on each other. Calculate the ratio of these two forces, and explain briefly:

$$\frac{F_2}{F_1} =$$

## Quiz for Chapter 2

A metal ball hanging from a thread is negatively charged. When a neutral glass ball is hung from a thread near the metal ball, the two balls attract each other.

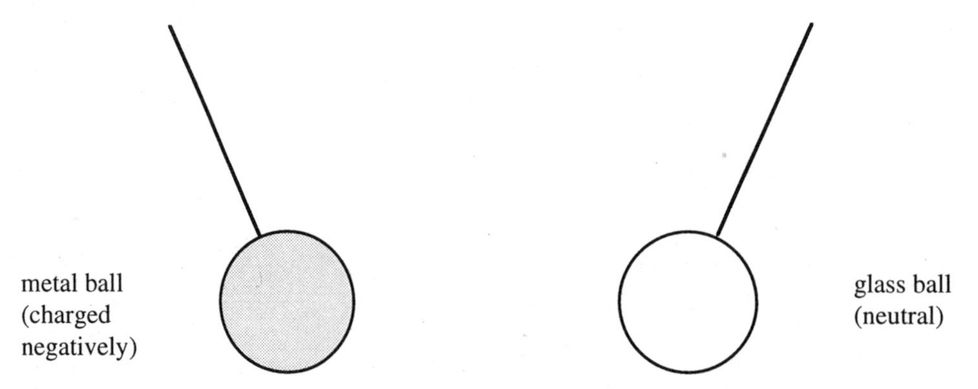


Explain this attraction briefly but in detail. Include careful, detailed diagrams.

## Quiz for Chapter 3

A lithium nucleus (3 protons and 4 neutrons, with mass = $\dfrac{7\times10^{-3}\,\text{kg / mole}}{6\times10^{23}\,\text{atoms / mole}}$) accelerates to the right due to electric forces, and the initial magnitude of the acceleration is $3\times10^{13}$ meters per second per second.

(a, 4 pts) What is the direction of the electric field that acts on the lithium nucleus?

(b, 8 pts) What is the magnitude of the electric field that acts on the lithium nucleus? Be quantitative (that is, give a number). Explain briefly.

(c, 8 pts) If this acceleration is due solely to a single helium nucleus (2 protons and 2 neutrons), where is the helium nucleus initially located? Be quantitative (that is, give a number). Explain briefly.

## Quiz for Chapter 4

A thin plastic ring of radius R is rubbed all over with a wool cloth, and acquires a charge -Q which is uniformly spread all over its surface. The ring is laid flat on a table. Using the four-step procedure presented in the workbook, find the *magnitude and direction* of the electric field due to the ring at a location a distance z above center of the ring. Show all of your work.

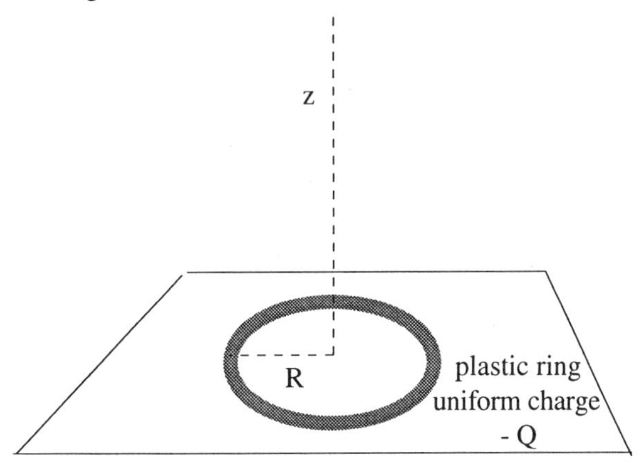


## Quiz for Chapter 6

In the circuit below, a wire of uniform thickness is connected to a single battery. The metal wire is 45 cm long, and has a cross-sectional area of $3.5\times10^{-8}$ m$^2$. The mobility of the metal is $5\times10^{-5}$ (m/s)/(N/C), and there are $8\times10^{28}$ free electrons per cubic meter in the metal. The battery does $2.4\times10^{-19}$ joules of work to transport an electron from its positive to its negative pole.

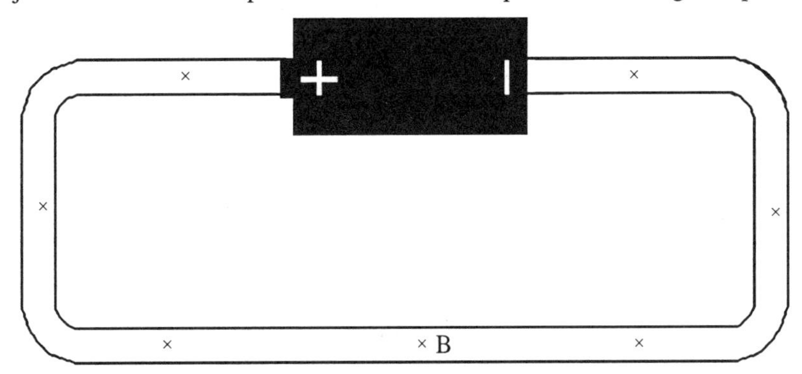


a) On the diagram carefully draw plusses and minuses to show the steady state surface charge distribution. Indicate higher charge density by placing +'s or -'s closer together, lower charge density by placing them far apart.

b) At the locations labeled by ×'s draw arrows showing the direction and relative magnitude of the steady state electric field.

c) Calculate quantitatively (with a numerical result) the electron current i flowing through the wire in the steady state. Show all your work, and explain your calculation (briefly). Include units!

d) Calculate quantitatively (with a numerical result) the magnitude of the electric field at location B inside the wire in the steady state. Show all your work, and explain your calculation (briefly). Include units!

| $i = nAv$ | $v = uE$ | $e = 1.6\times10^{-19}$ C | $\frac{1}{4\pi\varepsilon_0} = 9\times10^9 \frac{N-m^2}{C^2}$ |
|---|---|---|---|
| Current conservation: # of electrons/sec entering a region = # of electrons/sec leaving that region | | | |
| Energy conservation: $F_{NC}s = eE_1L_1 + eE_2L_2 + \ldots$ | | | |

## Quiz for Chapter 7

Two circuits (labeled 1 and 2) have different capacitors but the same batteries and long bulbs. The capacitors in circuit 1 and circuit 2 are identical except that the capacitor in circuit 2 was constructed with its plates closer together. Both capacitors have air between their plates. The capacitors are initially uncharged. In each circuit the batteries are connected for a time short compared to the time required to reach static equilibrium and then disconnected.

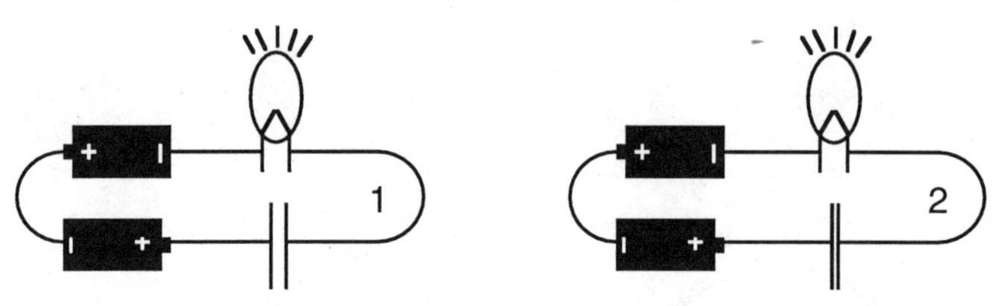


In which circuit (1 or 2) does the capacitor now have more charge? Explain your answer in detail.

## Quiz for Chapter 8

In a television picture tube electrons are boiled out of a very hot metal filament placed near a negative metal plate. These electrons start out nearly at rest and are accelerated toward a positive metal plate. They pass through a hole in the positive plate on their way toward the picture screen. The electric field is approximately uniform between the plates. The high-voltage supply in the television set maintains a potential difference of 20000 volts between the two plates. *As usual, show all your work, including calculations!*

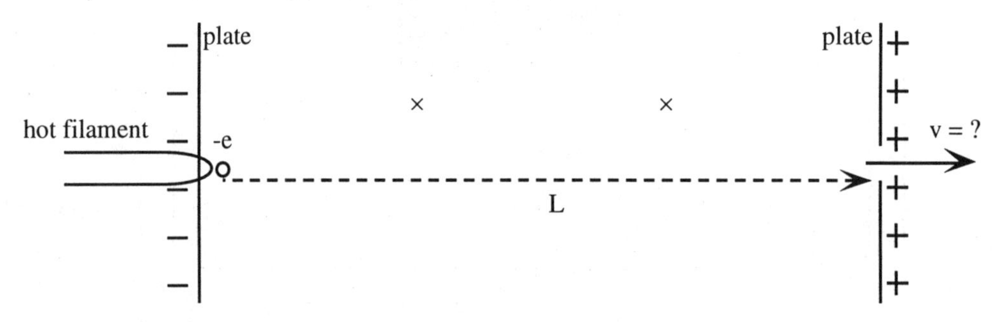


(a, 3 pts) Draw the electric field at the locations marked ×.

(b, 3 pts) Which plate is at a higher potential, the positive plate or the negative plate? Explain briefly.

(c, 7 pts) What is the speed v of the electrons as they pass through the hole in the positive plate? (Assume L is *not* known.)

(d, 7 pts) If the plates are a distance L = 8 mm apart, what is the magnitude of the electric field?

---

$V_f - V_i = \Delta V = \frac{\Delta PE}{q} = -\int_i^f \vec{E} \bullet d\vec{l}$ $\quad \frac{1}{4\pi\varepsilon_0} = 9 \times 10^9 \frac{N-m^2}{C^2}$ $\quad \Delta V_{\text{point charge}} = \frac{Q}{4\pi\varepsilon_0}\left[\frac{1}{r_f} - \frac{1}{r_i}\right]$

$m_{electron} = 9 \times 10^{-31}$ kg $\quad e = 1.6 \times 10^{-19}$ C

## Quiz for Chapter 11

This battery outputs a short-circuit current of 3.2 amperes. A portion of the wire that is nearly straight is aligned north-south and held a short distance above a compass as shown.

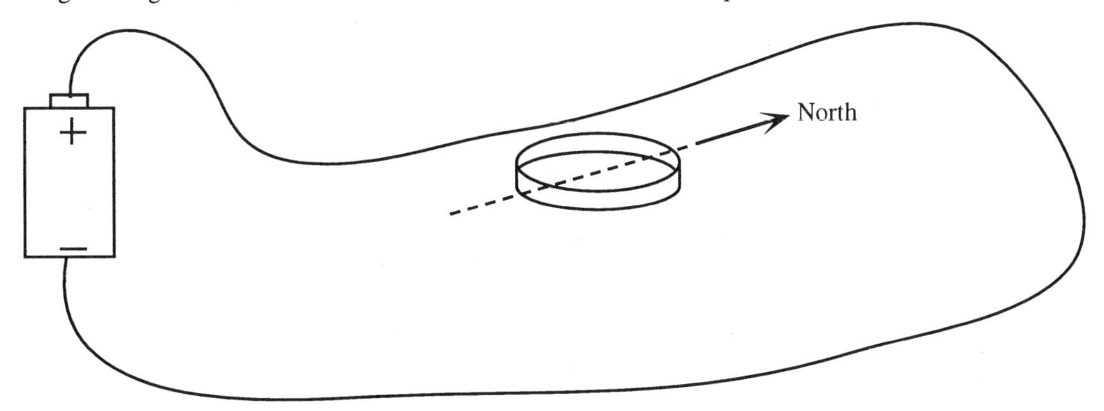


(a, 5 pts) The compass needle deflects 35 degrees away from north. Draw the compass needle direction on the diagram.

(b, 15 pts) How high above the compass is the wire? Explain briefly.

---

$\vec{B} = \frac{\mu_0}{4\pi}\frac{q\vec{v}\times\hat{r}}{r^2}$ $\Delta\vec{B} = \frac{\mu_0}{4\pi}\frac{I\Delta\vec{l}\times\hat{r}}{r^2}$ $B_{wire} \approx \frac{\mu_0}{4\pi}\frac{2I}{x}$ near a wire $\frac{\mu_0}{4\pi} = 10^{-7}\frac{\text{tesla} - \text{m}}{\text{ampere}}$

$B_{circular\ loop} \approx \frac{\mu_0}{4\pi}\frac{2\pi a^2 I}{x^3}$ along axis, far from loop $B_{earth\ (hor.)} = 2\times10^{-5}$ tesla

## Quiz for Chapter 12

An electron is moving at constant speed in a circle, perpendicular to a uniform magnetic field, making one clockwise revolution every 1.8 microseconds.

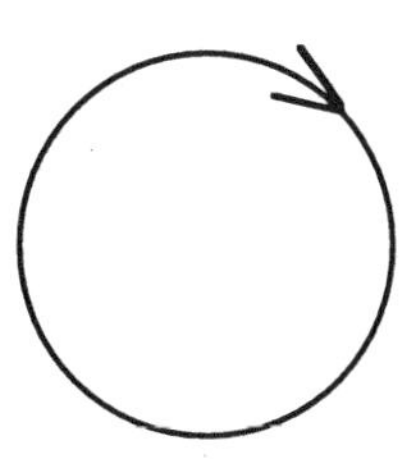

(a, 8 pts) What is the direction of the magnetic field? Explain briefly.

(b, 12 pts) What is the numerical magnitude B of the magnetic field?

---

$\vec{B} = \frac{\mu_0}{4\pi}\frac{q\vec{v}\times\hat{r}}{r^2}$ $\vec{F} = q\vec{E} + q\vec{v}\times\vec{B}$ $\frac{\mu_0}{4\pi} = 10^{-7}\frac{\text{tesla} - \text{m}}{\text{ampere}}$ $a_{radial} = \frac{v^2}{r}$ $m_{electron} = 9\times10^{-31}$ kg

## Quiz for Chapter 14

An electron is initially at rest. At time t = 0 it is accelerated upward with an acceleration of $10^{18}$ m/s$^2$ (this large acceleration is possible because the electron has a very small mass), and it continues accelerating. We make observations at location A, 15 meters from the electron.

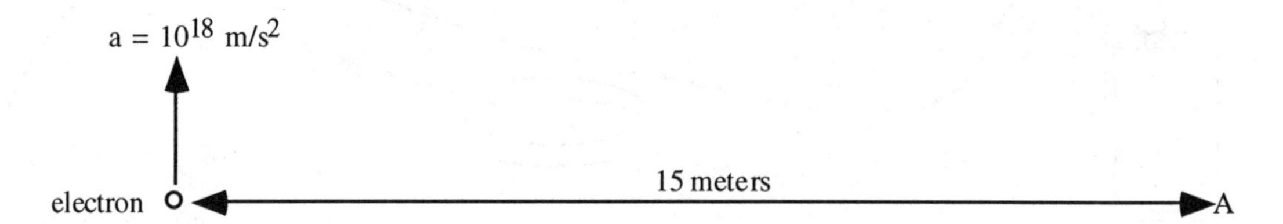


(a, 2 pts) At t = 1 ns ($10^{-9}$ sec), what is the magnitude and direction of the electric field at location A due to the electron? (Give numerical value.)

(b, 3 pts) When does the electric field at location A change?

(c, 5 pts) What is the initial direction of the new electric field?

(d, 5 pts) What is the initial magnitude of the new electric field? (Give numerical value.)

(e, 5 pts) What is the direction of the initial *magnetic* force on a positive charge at location A? Explain with a diagram.

---

$\vec{E}_{radiative} = \frac{1}{4\pi\varepsilon_0}\frac{-q\vec{a}_\perp}{c^2 r}$ direction of propagation is $\vec{E}\times\vec{B}$. $\vec{E} = \frac{1}{4\pi\varepsilon_0}\frac{q}{r^2}\hat{r}$

$\frac{1}{4\pi\varepsilon_0} = 9\times10^9 \text{N}-\text{m}^2/\text{C}^2$ $e = 1.6\times10^{-19}$ C $c = 3\times10^8$ m/s

## Exam 1 covering Chapters 1 through 4

- **Read all problems carefully before attempting to solve them.**
- **Correct answers without adequate explanation *will be counted wrong*.**
- **Incorrect explanations mixed in with correct explanations *will be counted wrong*.**
- **Make explanations complete but brief. Do not write a lot of prose.**
- **Include diagrams where needed.**

***There is a tear-off formula sheet on the back.***

**Problem 1 (30 pts):** A lightweight metal ball hangs from a thread, to the right of a plastic rod. Both the rod and ball are initially uncharged. You rub the left end of the plastic rod with wool, depositing charged molecular fragments whose total (negative) charge is that of $10^9$ electrons. After moving the wool far away, you observe that the ball moves a short distance toward the rod as shown here:

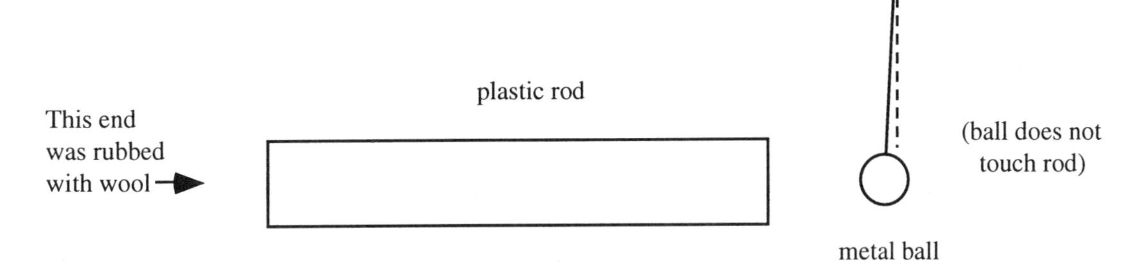


**(a, 15 pts)** Explain briefly. Show all excess charged particles, polarization, etc. *clearly* in the diagram. Make it clear whether charged particles that you show are on the surface of an object or inside it.

You perform a similar experiment with a metal rod. You touch the left end of the rod with a charged metal object, depositing $10^9$ excess electrons on the left end (and then you remove the object). You see the ball deflect more than it did with the plastic rod (part "a" above).

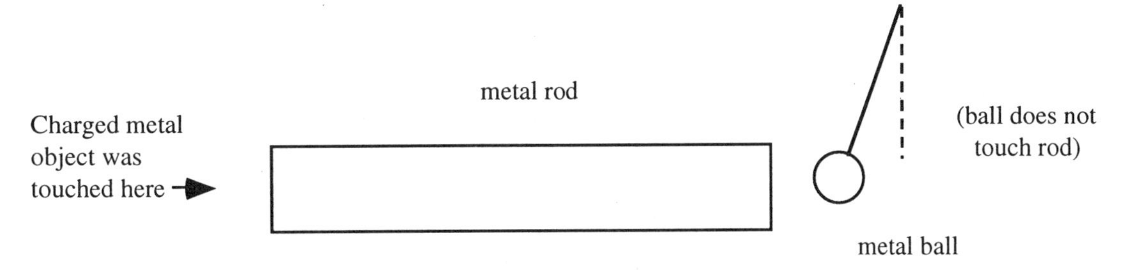


**(b, 15 pts)** Explain briefly. Show all excess charged particles, polarization, etc. *clearly* in the diagram. Make it clear whether charged particles that you show are on the surface of an object or inside it.

**Problem 2 (35 pts):** A thin, hollow spherical plastic shell of radius R carries a uniformly distributed negative charge -Q. (A slice through the plastic shell is shown.) To the left of the spherical shell are four charges packed closely together as shown (the distance "a" is shown greatly enlarged for clarity). The distance from the center of the four charges to the center of the plastic shell is L, which is much larger than a (L >> a).

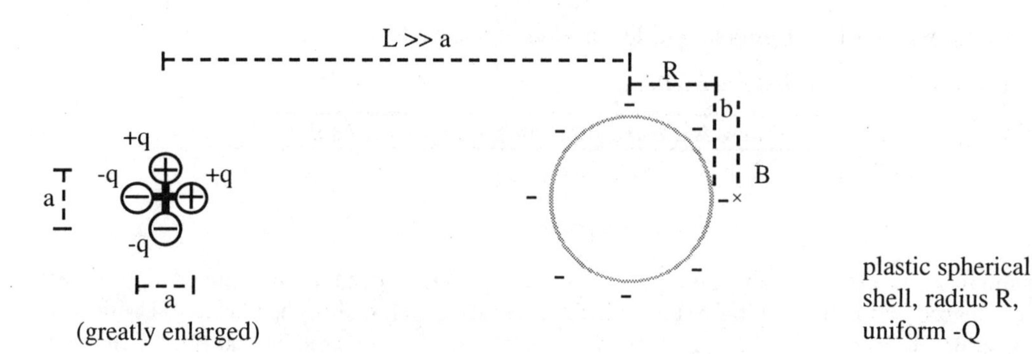


**(a, 16 pts):** Calculate the **x- and y-components** of the electric field at location B, a distance b to the right of the outer surface of the plastic shell. Explain briefly, *including showing the electric field on the diagram.* Your results *must not contain* any symbols other than the given quantities R, Q, q, a, L, and b (and fundamental constants). Don't try to simplify the final algebraic results except for taking into account the fact that L >> a.

**(b, 3 pts):** What simplifying assumption did you have to make in part "a"?

**(c, 16 pts):** The plastic shell is removed and replaced by an uncharged metal ball.

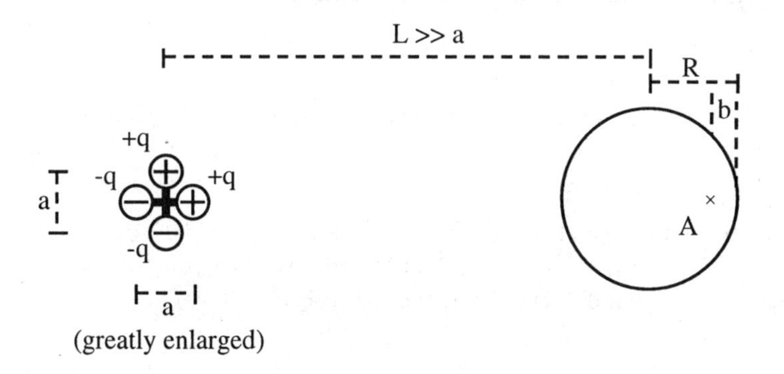


solid metal
ball, radius R,
zero net charge

(1) At location A inside the metal ball, a distance b to the left of the outer surface of the ball, *accurately* draw and label the electric field $\vec{E}_{ball}$ due to the ball charges and the electric field $\vec{E}_4$ of the four charges.

(2) Show the distribution of ball charges.

(3) Calculate the **x- and y-components** of the net electric field at location A.

Explain briefly.

**Problem 3 (35 pts):** A very thin plastic rod of length L is rubbed with cloth and becomes uniformly charged negatively, with net charge -Q.

**(a, 30 pts):** Using the standard four-step procedure, calculate the ***magnitude and direction of the electric field*** at location ×, a distance b to the right of the end of the rod. If you introduce new geometrical quantities, ***be sure to define them with labels on the diagram***. Your final result for the magnitude of the field must be expressed in terms of L, Q, and b.

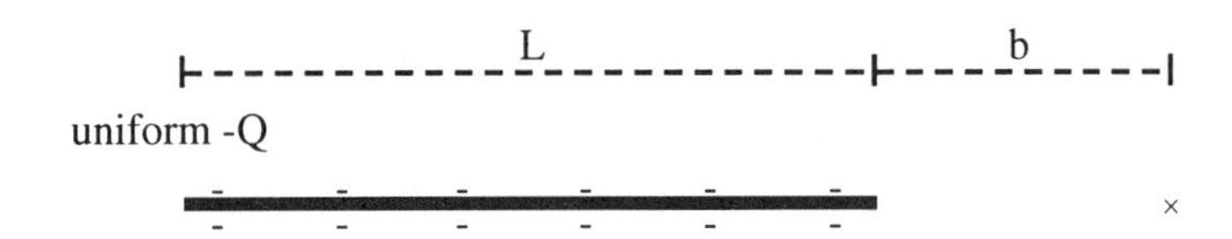


**(b, 5 pts):** Suppose the rod is 15 cm long, and there are $10^{11}$ excess electrons (or singly-charged negative ions) on the rod. An electron is placed 3 cm to the right of the right end of the rod. What is the magnitude of the force on the electron?

F =

**Formula sheet for Exam 1**

$\frac{1}{4\pi\varepsilon_0} = 9\times10^9$ N-m$^2$/C$^2$

$e = 1.6\times10^{-19}$ coulomb

$\varepsilon_0 = 9\times10^{-12}$ C$^2$/N-m$^2$

$g = 9.8$ N/kg

$$m_{\text{hydrogen atom}} = \frac{(1\times10^{-3}\text{ kilogram/mole})}{(6\times10^{23}\text{ atoms/mole})} = 1.7\times10^{-27}\text{ kilogram}$$

$m_{\text{electron}} = 9\times10^{-31}$ kg

$$E_{\text{point or sphere}} = \frac{1}{4\pi\varepsilon_0}\frac{Q}{d^2}$$

$E_{\text{breakdown in air}} \approx 3\times10^6$ N/C

$$E_{\text{dipole}} \approx \frac{1}{4\pi\varepsilon_0}\left[\frac{2qs}{d^3}\right]$$ at a distance d from the dipole midpoint, along the axis of the dipole

$$E_{\text{dipole}} \approx \frac{1}{4\pi\varepsilon_0}\left[\frac{qs}{d^3}\right]$$ at a distance d from the dipole midpoint, perpendicular to axis of the dipole

dipole moment $p = qs = \alpha E$, where $\alpha$ is the "polarizability"

$$E_{\text{rod}} = \frac{Q}{4\pi\varepsilon_0}\left[\frac{1}{x\sqrt{x^2+(L/2)^2}}\right]$$, perp. to rod

$$E_{\text{rod}} \approx \frac{1}{4\pi\varepsilon_0}\frac{2Q/L}{x}$$, if $x << L$

$$E_{\text{ring}} = \frac{1}{4\pi\varepsilon_0}\frac{qx}{(x^2+r^2)^{3/2}}$$

area of circle = $\pi r^2$; circumference = $2\pi r$; area of cylinder = $2\pi rL$

$$E_{\text{disk}} = \frac{Q/A}{2\varepsilon_0}\left[1-\frac{x}{\left(x^2+R^2\right)^{1/2}}\right]$$

$$E_{\text{disk}} \approx \frac{Q/A}{2\varepsilon_0}$$, if $x << R$

$$E_{\text{capacitor}} \approx \frac{Q/A}{\varepsilon_0}$$ (+Q disk and -Q disk)

$$E_{\text{fringe}} \approx \frac{Q/A}{\varepsilon_0}\left(\frac{s}{2R}\right)$$ just outside the capacitor

## Exam 2 covering Chapters 5 through 9

- **Read all problems carefully before attempting to solve them.**
- **Correct answers without adequate explanation *will be counted wrong.***
- **Incorrect explanations mixed in with correct explanations *will be counted wrong.***
- **Make explanations complete but brief. Do not write a lot of prose.**
- **Include diagrams where needed.**

| *There is a tear-off formula sheet on the back.* |
| --- |
| *Use only these formulas! Derive anything else.* |

**Problem 1 (40 pts).** A circuit contains an ideal mechanical battery which exerts a non-Coulomb force $F_{NC}$ to move electrons through the battery (with negligible internal resistance). The end plates of the battery are very large compared to the distance 0.3L between the plates (plates not drawn to scale). Two thin nichrome wires of length L and cross-sectional area A connect the battery to a thick nichrome wire of length 0.6L and cross-sectional area 4A. The mobility of the nichrome is u, and there are n mobile electrons per cubic meter in the nichrome.

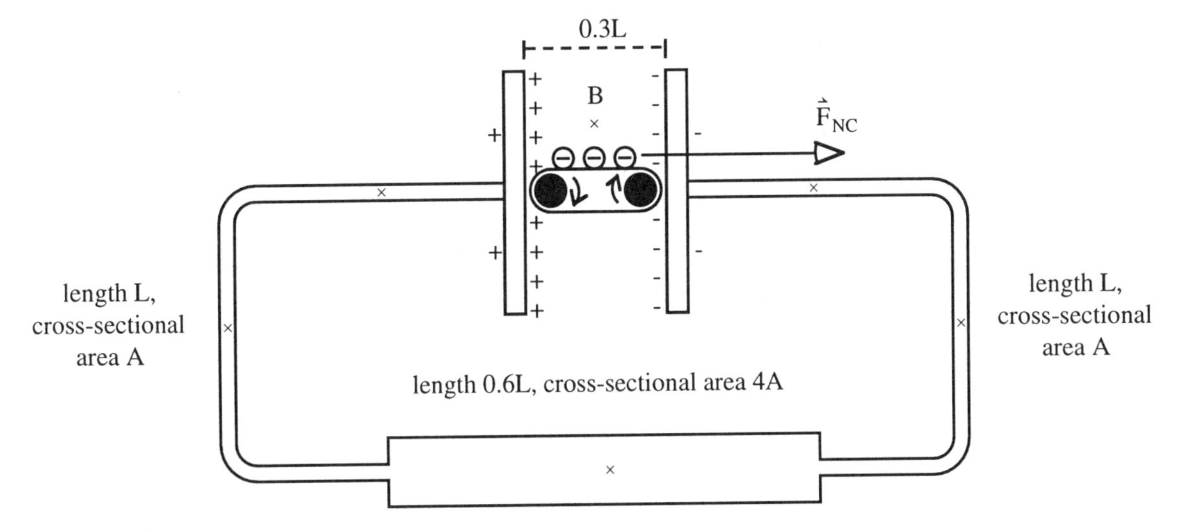


**(a, 9 pts)** Show the electric field at the ***six*** locations marked with × (including location B between the plates). Pay attention to the relative magnitudes of the ***six*** vectors that you draw.

**(b, 6 pts)** Show the approximate distribution of charge on the surface of the nichrome wires. Make sure that your distribution is compatible with the electric fields that you drew in part (a).

**(c, 20 pts)** Calculate the number of electrons that leave the battery every second, in terms of the given quantities L, A, n, u, and $F_{NC}$ (and fundamental constants). Be sure to show all of your work.

**(d, 5 pts)** Calculate the magnitude of the electric field between the plates of the battery (location B).

**Problem 2 (40 pts).** A long iron slab of width w and height h emerges from a furnace. Because the end of the slab near the furnace is hot, and the other end is cold, the electron mobility increases significantly with the distance x (see diagram): $u = u_1 + kx$, where $u_1$ is the mobility of the iron at the hot end of the slab. There are n iron atoms per cubic meter, and each atom contributes one electron to the sea of mobile electrons (we can neglect the small thermal expansion of the iron). A steady-state conventional current runs through the slab from the hot end toward the cold end, and an ammeter (not shown) measures the current to have a magnitude I in amperes. A voltmeter is connected to two locations a distance d apart, as shown.

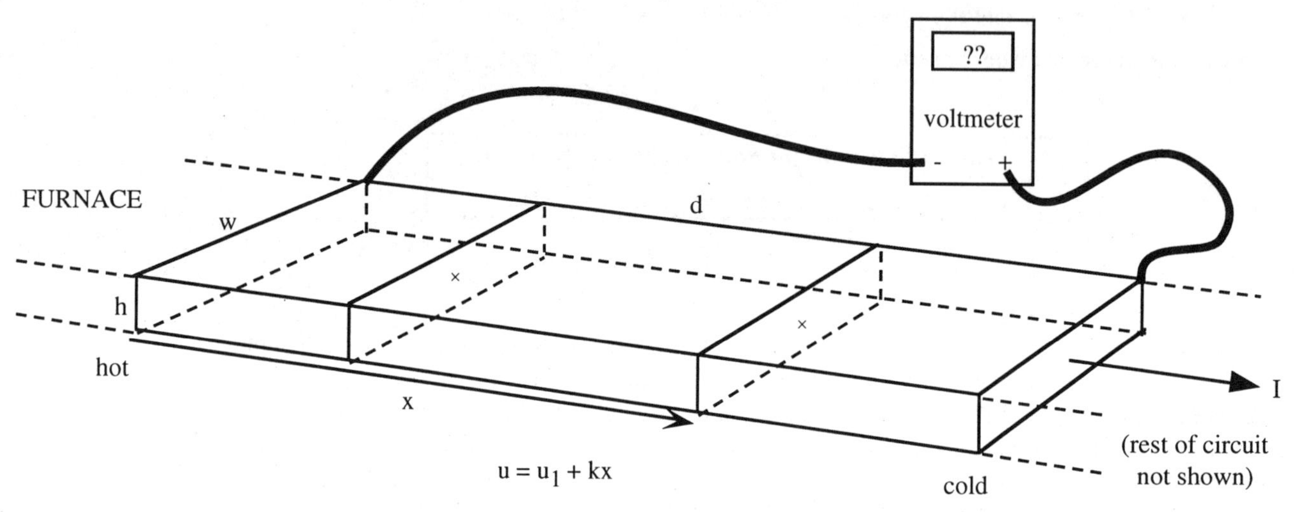


**(a, 5 pts)** Show the electric field inside the slab at the two locations marked with ×. Pay attention to the relative magnitudes of the two vectors that you draw.

**(b, 10 pts)** Explain why the magnitude of the electric field is different at these two locations:

**(c, 5 pts)** At a distance x from the left voltmeter connection, what is the magnitude of the electric field in terms of x and the given quantities w, h, d, $u_1$, k, n, and I (and fundamental constants)?

**(d, 5 pts)** What is the sign of the potential difference displayed on the voltmeter? Explain briefly.

**(e, 10 pts)** In terms of the given quantities w, h, d, $u_1$, k, n, and I (and fundamental constants), what is the magnitude of the voltmeter reading? If you can't fully evaluate this, go as far as you can. Check your work.

**(f, 5 pts)** What is the resistance of this length of the iron slab?

Check that your results for parts c, e, and f contain ONLY the given quantities!

**Problem 3 (20 pts):** A capacitor with a slab of glass between the plates is connected to a battery by nichrome wires and allowed to charge completely. Then the slab of glass is removed. Describe and explain what happens. If you give a direction for a current, state whether you are describing *electron current* or *conventional current*.

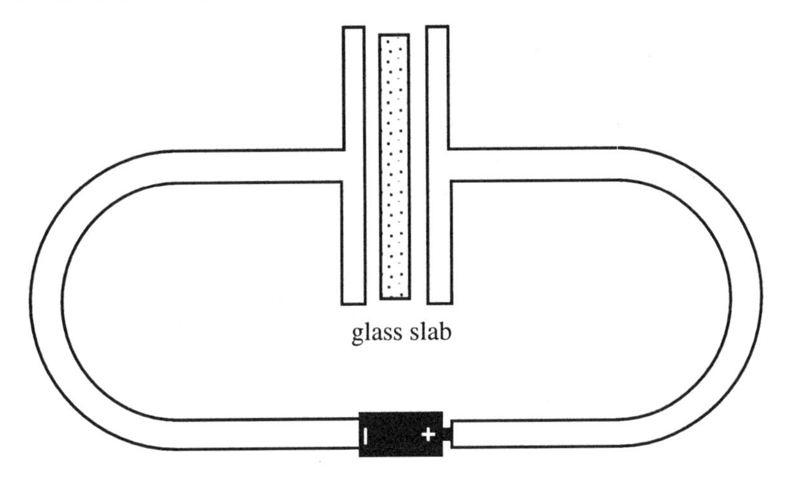

**Formula sheet for Exam 2**

$\frac{1}{4\pi\varepsilon_0} = 9\times10^9$ N-m$^2$/C$^2$

$e = 1.6\times10^{-19}$ coulomb

$\varepsilon_0 = 9\times10^{-12}$ C$^2$/N-m$^2$

$g = 9.8$ N/kg

$m_{\text{hydrogen atom}} = \frac{(1\times10^{-3}\text{ kilogram/mole})}{(6\times10^{23}\text{ atoms/mole})}$

$m_{\text{electron}} = 9\times10^{-31}$ kg

$E_{\text{point charge}} = \frac{1}{4\pi\varepsilon_0}\frac{Q}{r^2}$

$E_{\text{breakdown in air}} \approx 3\times10^6$ N/C

Very near flat sheet of charge: $E \approx \frac{Q/A}{2\varepsilon_0}$

$\Delta V \equiv \frac{\Delta PE}{q} = -\int_i^f \vec{E}\bullet d\vec{l}$

Two-disk capacitor, at center (for d << R):

$E = -\frac{\Delta V}{\Delta L}$

inside: $E \approx \frac{Q/A}{\varepsilon_0}$

just outside: $E \approx \frac{Q/A}{\varepsilon_0}\frac{s}{2R}$

$Q = C|\Delta V|$

Dipole with charge q and -q

separated by s along x:

$E_x = \frac{1}{4\pi\varepsilon_0}\frac{2sq}{x^3}$ along x-axis

$E_x = -\frac{1}{4\pi\varepsilon_0}\frac{sq}{y^3}$ along y-axis

$\Delta V_{\text{round trip}} = 0$

$\Sigma(I) = 0$

power = $I\Delta V$

$F_{NC}d = \Sigma(eEL)$

$E \rightarrow E/K$ inside insulator

$i = \#/\text{sec} = nAv$; $v = uE$

$R = \frac{L}{\sigma A}$

$J = I/A = qnv = \sigma E$

$I = \frac{|\Delta V|}{R}$

$\Delta V_{\text{battery}} = \text{emf} - r_{\text{internal}}I$

$\text{emf} = \frac{F_{NC}s}{e}$

$Q = Q_{\text{final}}\left[1 - e^{-\frac{t}{RC}}\right]$ (charging)

$Q = Q_{\text{initial}}e^{-\frac{t}{RC}}$ (discharging)

## Exam 3 covering Chapters 10 through 13

- **Read all problems carefully before attempting to solve them.**
- **Correct answers without adequate explanation** ***will be counted wrong.***
- **Incorrect explanations mixed in with correct explanations** ***will be counted wrong.***
- **Make explanations complete but brief. Do not write a lot of prose.**
- **Include diagrams where needed.**

*If you refer to a right-hand rule, provide sufficient context to make it clear which of the many right-hand rules you mean!*

*There is a tear-off formula sheet on the back.*

**Problem 1 (35 pts):** Two straight wires carrying conventional current I are connected by a three-quarter-circular arc of radius $r_1$ and a one-quarter-circular arc of radius $r_2$. Electric fields are also present in this region, due to charges not shown on the drawing.

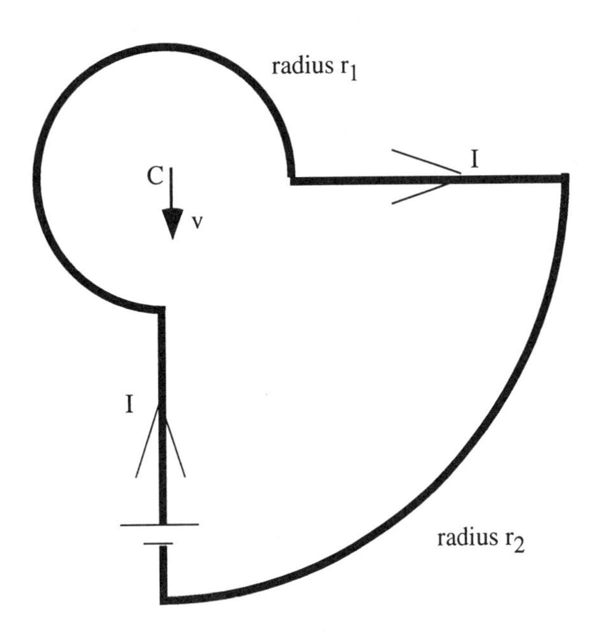


An electron is moving down with speed v as it passes through the center C of the arc, and at that instant the net electric and magnetic force on the electron is zero.

---

**(a, 5 pts)** What is the direction of the magnetic field at the center C, due to the current I? Explain briefly.

**(b, 8 pts)** What is the direction of the electric field at the center C? Explain your reasoning clearly.

**(c, 15 pts)** What is the magnitude of the magnetic field at the center C, due to the current I? *Start from fundamental principles. Do NOT use any formulas not given on the formula page.*

**(d, 7 pts)** What is the magnitude of the electric field at the center C? Explain briefly.

**Problem 2 (35 pts):** A long straight wire carries a conventional current I to the right as shown, and this current is decreasing: I = p - kt, where t is the time in seconds, and p and k are positive constants. At a distance y from the wire is a thin rectangular coil of width w and height h containing 3 turns of wire with total resistance R. This is a top view—the coil lies on a frictionless table. The coil is initially at rest.

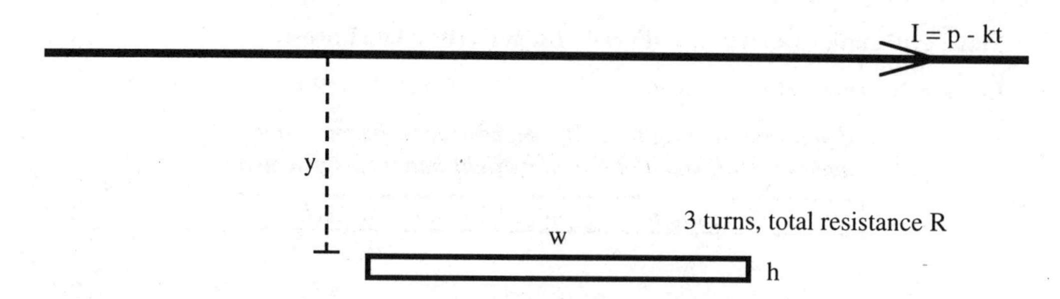


What are the initial **magnitude and direction of the non-zero net force** that is acting on the coil? Explain in detail. If you make any simplifying assumptions, state clearly what they are, but bear in mind that the net force is *not* zero.

**Problem 3 (15 pts):** You hold your magnet perpendicular to a bar of copper which is connected to a battery as shown:

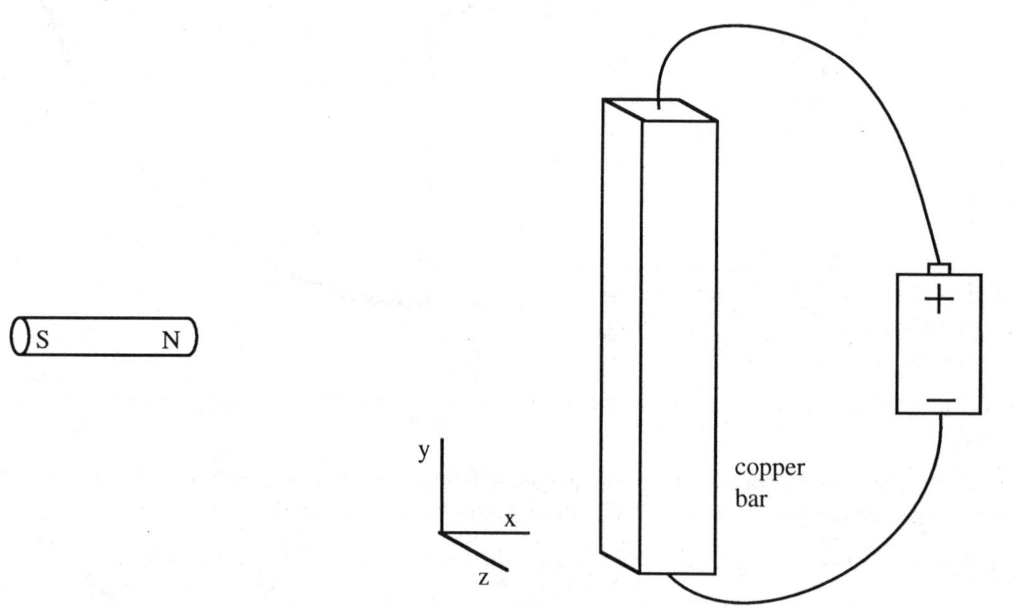


Describe the directions of the components of the electric field at the center of the copper bar, both parallel to the bar and perpendicular to the bar. Explain carefully.

**Problem 4 (15 pts):** A long plastic rod, of length L and diameter d, is rubbed all over with a wool cloth. It acquires a net negative charge of -Q, uniformly distributed over the entire curved surface of the rod. Using the cylindrical Gaussian surface shown on the diagram, find the magnitude and direction of the electric field a distance z to the right of the midpoint of the rod (along a line perpendicular to the rod).

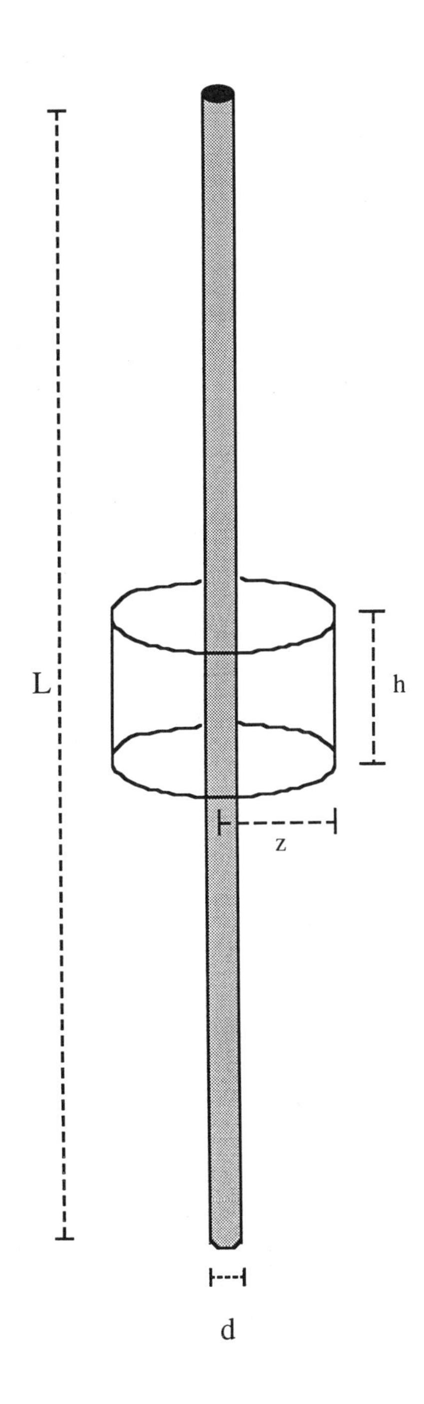

**Formula sheet for Exam 3**

$\frac{1}{4\pi\varepsilon_0} = 9\times10^{9}\ \text{N-m}^2/\text{C}^2$

$e = 1.6\times10^{-19}$ coulomb

$\varepsilon_0 = 9\times10^{-12}\ \text{C}^2/\text{N-m}^2$

$g = 9.8$ N/kg

$\frac{\mu_0}{4\pi} = 10^{-7}\ \frac{\text{tesla-m-sec}}{\text{C}}$

$m_{electron} = 9\times10^{-31}$ kg

circle circumference = $2\pi r$

circle area = $\pi r^2$

$$\oint_{\text{closed surface}} \vec{E}\bullet\hat{n}dA = \frac{\sum q_{inside}}{\varepsilon_0}$$

$$\oint_{\text{closed surface}} \vec{B}\bullet\hat{n}dA = 0 \text{ (Gauss's Law for magnetism)}$$

$$\vec{F} = q\vec{E} + q\vec{v}\times\vec{B}$$

$$d\vec{F} = Id\vec{l}\times\vec{B}$$

$$\vec{B} = \frac{\mu_0}{4\pi}\frac{q\vec{v}\times\hat{r}}{r^2}$$

$$d\vec{B} = \frac{\mu_0}{4\pi}\frac{Id\vec{l}\times\hat{r}}{r^2}$$

$$\text{emf}_{\text{along bounding path}} = \oint_{\text{along bounding path}} \vec{E}\bullet d\vec{l} = -N\frac{d}{dt}\left(\int_{\text{open surface}} \vec{B}\bullet\hat{n}dA\right)$$

$B_{wire} \approx \frac{\mu_0}{4\pi}\frac{2I}{x}$ a distance x from a long straight wire

$B \approx \frac{\mu_0}{4\pi}\frac{2\mu}{x^3}$, with $\mu = IA$, along the axis a distance x, far from a current loop

$B_{solenoid} \approx \frac{\mu_0 NI}{L}$ inside a long solenoid with N turns (radius of coil << length L)

## Final exam covering Chapters 1 through 14

- **Read all problems carefully before attempting to solve them.**
- **Correct answers without adequate explanation *will be counted wrong*.**
- **Incorrect explanations mixed in with correct explanations *will be counted wrong*.**
- **Make explanations complete but brief. Do not write a lot of prose.**
- **Include diagrams where needed.**

***The last two pages are tear-off lists of useful information.***

**Prob. 1 (80 pts):**

**(a, 10 pts)** A bar magnet is held vertically above a horizontal metal ring. The north pole of the magnet is at the top. If the magnet is lifted straight up, draw on the diagram the direction that current runs in the metal ring. Explain briefly.

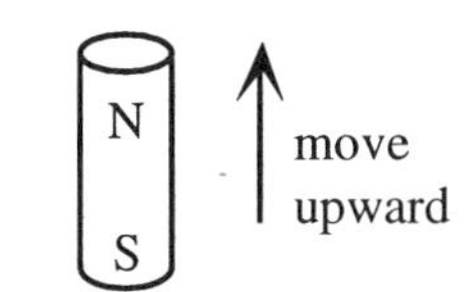

**(b, 10 pts)** When you charged your capacitor through a round bulb, it took about 5 seconds to charge. When you charged your capacitor through a long bulb, it took about 15 seconds to charge. Compare the final amount of charge on the positive plate of the capacitor in the two cases, and explain briefly.

**(c, 10 pts)** The magnetic field in a region is horizontal and was measured to have the values shown on the surface of a cylinder. Why should you suspect something is wrong with these measurements?

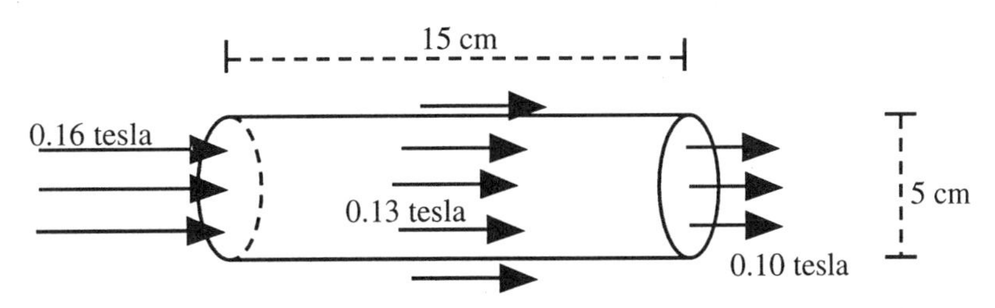


**(d, 10 pts)** Outline how you could measure the internal resistance of one of your flashlight batteries, using only the equipment in your experiment kit (tape, circuit apparatus, magnetism apparatus).

**(e, 10 pts)** At t = 0, an electron starts to accelerate downward, at a distance of 3 meters to the right. At what time do you first observe radiative electric and magnetic fields, and what are their directions?

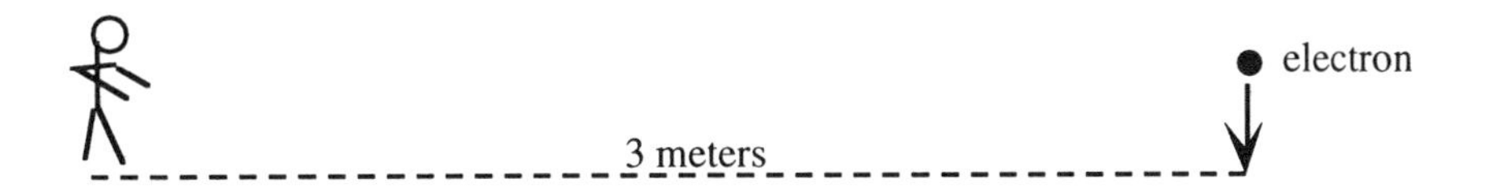

**(f, 10 pts)** You rub a plastic pen through your hair. Explain carefully but briefly why the pen attracts both pieces of aluminum foil and pieces of paper.

**(g, 10 pts)** The three circuits shown below consist of identical batteries and bulbs.

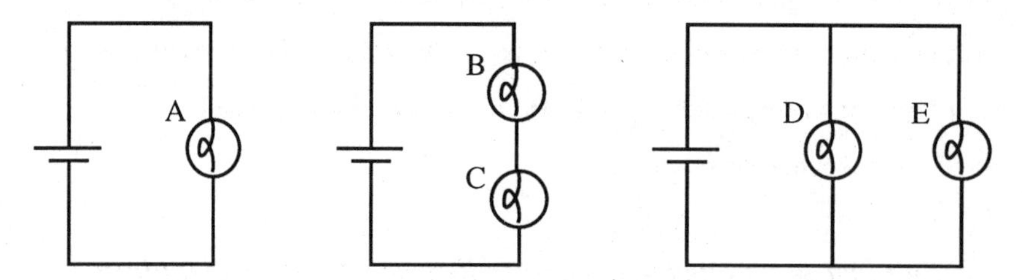


Rank the bulbs in order of brightness from brightest to dimmest. If two or more bulbs are equal in brightness, indicate this in your response. **Explain the reasoning you used to determine this ranking.**

**(h, 10 pts)** On a cloudless day, when you look away from the sun at the rest of the sky, the sky is bright and you cannot see the stars. On the moon, however, the sky is dark and you *can* see the stars even when the sun is visible. Explain briefly.

**Prob. 2 (20 pts):** In a lecture demonstration four 22-megohm resistors were in series with a 10000-volt power supply. The bare copper wires were suspended from insulating supports.

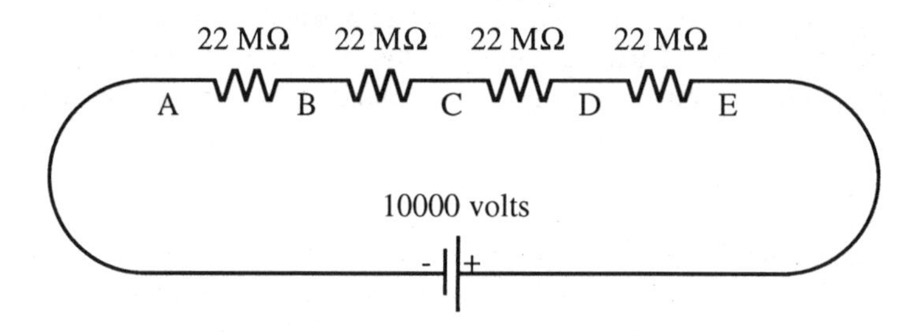


A small pith ball was painted with conducting paint and hung from a thread. The pith ball was touched to ground to discharge it. When the hanging pith ball was brought near location A, it swung toward A, touched the wire, then immediately jumped strongly away.

Explain this experimental observation in detail. Also predict and explain in detail what happens if the pith ball after being grounded is touched to locations B, C, D, or E.

**Problem 3 (35 pts):** A thin-walled hollow circular glass tube, open at both ends, has a radius R and length L. The axis of the tube lies along the x-axis, with the left end at the origin. It is charged uniformly with a charge +Q. Determine the electric field at a location on the x-axis, a distance d from the origin. Explain each step.

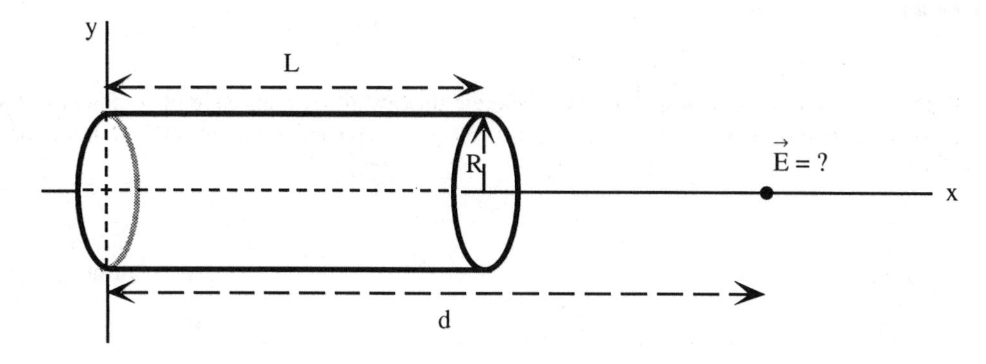

**Prob. 4 (20 pts):** Two large rectangular metal plates, 3 meters by 2 meters, are 2 mm apart. A thin sheet of glass 3 meters by 2 meters by 0.7 mm is placed on the lower metal plate. The dielectric constant of the glass is 5. A variable high-voltage power supply is connected to the metal plates.

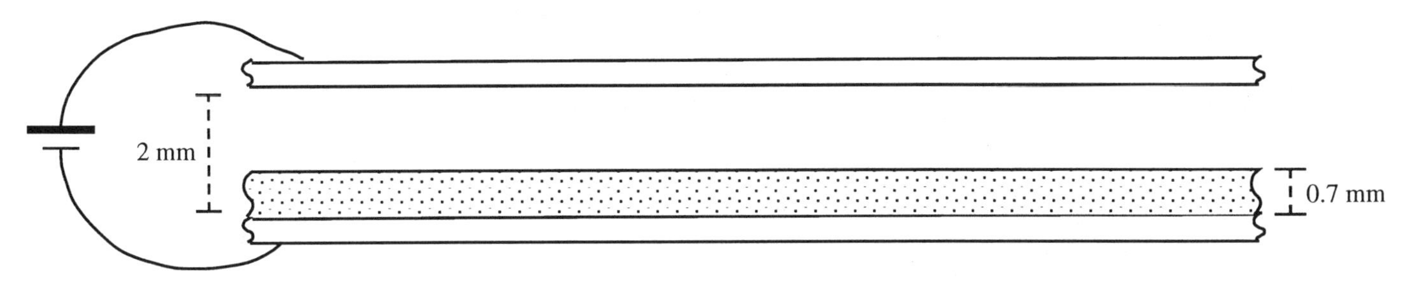


As the power supply voltage is slowly increased, at a certain voltage there is a spark in the air gap between the upper metal plate and the glass. Calculate the power-supply voltage at which this occurs.

**Problem 5 (20 pts):**

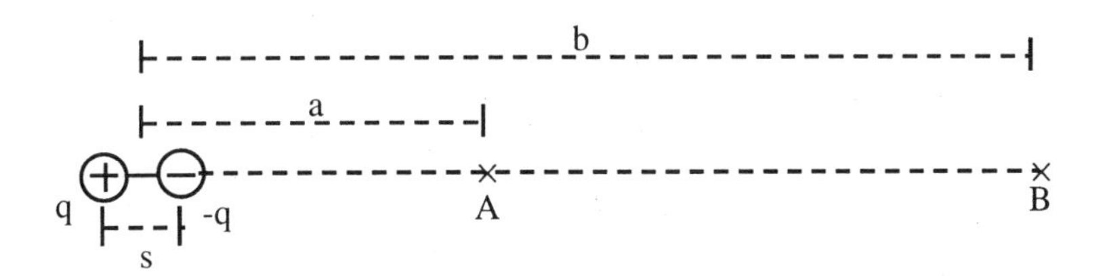


**(a, 15 pts)** Find $\Delta V = V_A - V_B$ along the axis of the dipole, where $s << a$. Include the correct sign. Explain carefully.

**(b, 5 pts)** What is the change in potential energy $\Delta PE$ in moving an electron from B to A?

**Prob. 6 (40 pts):** A bar of length L, width w, and thickness d is positioned a distance h underneath a long straight wire that carries a large but unknown steady current $I_{wire}$ (w is small compared to h). The material that the bar is made of is known to have n positive mobile charge carriers ("holes") per cubic meter, each of charge e, and these are the only mobile charge carriers in the bar. An ammeter measures the small current I passing through the bar.

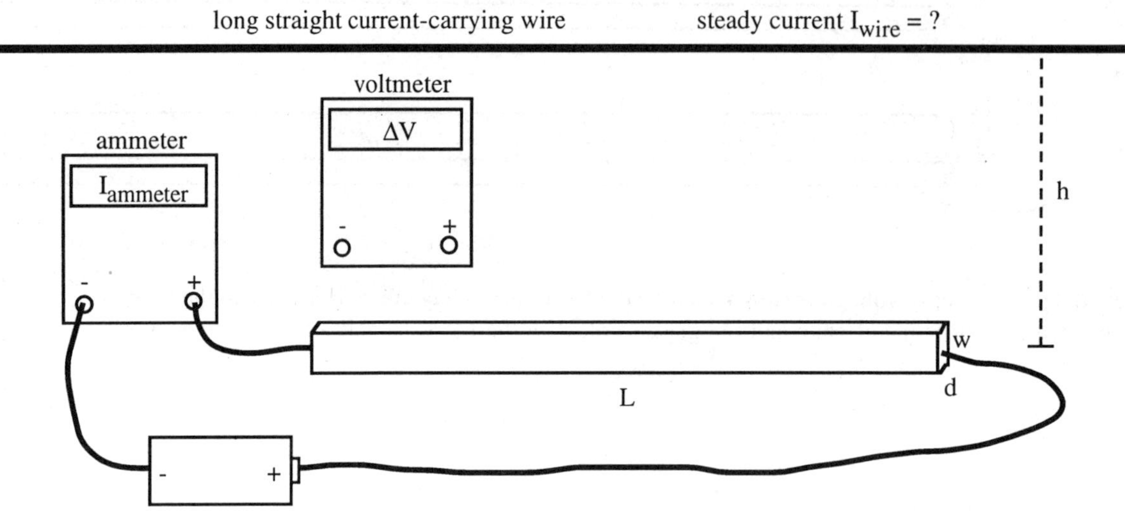


**(a, 20 pts)** Draw connecting wires from the voltmeter to the bar, showing clearly how the ends of the wires must be positioned on the bar in order to permit determining the amount of current in the long straight wire $I_{wire}$. If the voltmeter reading $\Delta V$ is positive, does the current in the long straight wire run to the left or to the right? Explain briefly.

**(b, 20 pts)** If the voltmeter reading is $\Delta V$, what is the current in the long straight wire, $I_{wire}$? Express your answer for the large current $I_{wire}$ only in terms of the known quantities: h, L, w (which is small compared to h), d, n, $I_{ammeter}$, $\Delta V$, and known physical constants such as e.

**Prob. 7 (40 pts):** The drawing shows a slide wire of length L = 0.6 m and mass m = 2.5 kg moving without friction along parallel conducting tracks with constant velocity $\vec{v}$ (v = 20 m/s). There is a battery ($emf_{bat}$ = 24 volts) and a resistor (R = 48 ohms) in the circuit, and a uniform magnetic field $\vec{B}$ pointing into the paper (B = 0.8 tesla). There is also an external force $\vec{F}_A$ applied to the wire to keep it moving at a constant velocity.

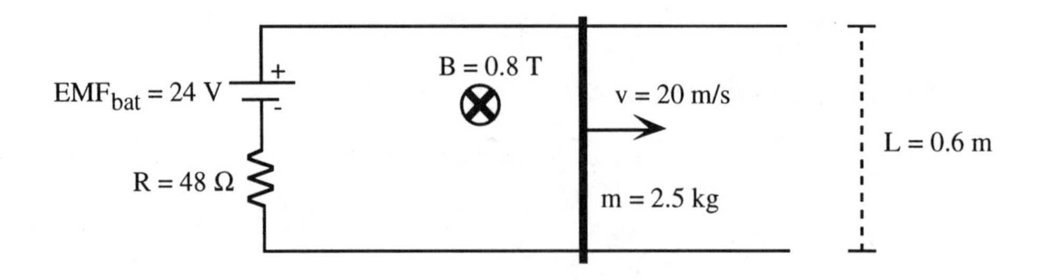


**(a, 30 pts)** Find the magnitude and direction of $\vec{F}_A$. Explain carefully:

Now the applied force $\vec{F}_A$ is abruptly removed from the slide wire at a time we shall call t = 0.

**(b, 10 pts)** Describe **qualitatively** what happens to the velocity of the wire as a function of time for t > 0, giving <u>cogent</u> but <u>brief</u> reasons for your statements:

**Formula sheets for Final Exam**

## CONSTANTS & GEOMETRY

$\frac{1}{4\pi\varepsilon_0} = 9\times10^9 \frac{\text{N-m}^2}{\text{C}^2}$ $\quad$ $\frac{\mu_0}{4\pi} = 10^{-7} \frac{\text{tesla-m-s}}{\text{C}}$ $\quad$ $c = 3\times10^8 \frac{\text{m}}{\text{s}}$

$\varepsilon_0 = 9\times10^{-12} \frac{\text{C}^2}{\text{N-m}^2}$ $\quad$ $e = 1.6\times10^{-19}$ C $\quad$ $g = 9.8 \frac{\text{N}}{\text{kg}}$

$E \approx 3\times10^6 \frac{\text{N}}{\text{C}}$ breakdown of air $\quad$ $B_{earth} \approx 2\times10^{-5}$ tesla

$m_{electron} = 9\times10^{-31}$ kg $\quad$ $m_{proton} = 1.7\times10^{-27}$ kg

circumference of circle $= 2\pi r$ $\quad$ area of circle $= \pi r^2$ $\quad$ surface area of sphere $= 4\pi r^2$

## ELECTRICITY

$\int_{\text{closed surface}} \vec{E}\cdot\hat{n}dA = \frac{\sum q_{inside}}{\varepsilon_0}$ $\quad$ $\vec{E}_{point} = \frac{1}{4\pi\varepsilon_0}\frac{q}{r^2}\hat{r}$ $\quad$ $E_{\text{inside insulator}} \rightarrow \frac{E}{K}$

$\Delta V = V_f - V_i = -\int_i^f \vec{E}\cdot d\vec{l}$ $\quad$ $E = -\frac{\Delta V}{\Delta l}$ $\quad$ $\text{Work} = q\Delta V = \Delta\left(\frac{1}{2}mv^2\right)$

$V - V_\infty = \frac{1}{4\pi\varepsilon_0}\frac{q}{r}$ $\quad$ Power $= I\Delta V$

$\#/_s = nAv$ $\quad$ $v = uE$ $\quad$ Ohmic: $I = \frac{\Delta V}{R}$

$\sum \Delta V_{\text{circuit loop}} = 0$ $\quad$ $\sum I_{\text{out of circuit node}} = 0$

## MAGNETISM

$\int_{\text{closed surface}} \vec{B}\cdot\hat{n}dA = 0$ $\quad$ $\int_{\text{closed path}} \vec{B}\cdot d\vec{l} = \mu_0\left[\sum I_{inside} + \varepsilon_0\frac{d\Phi_{electric}}{dt}\right]$ $\quad$ $\Phi_{electric} = \int_{\text{bounded surface}} \vec{E}\cdot\hat{n}dA$

$\vec{B} = \frac{\mu_0}{4\pi}\frac{q\vec{v}\times\hat{r}}{r^2}$ $\quad$ $d\vec{B} = \frac{\mu_0}{4\pi}\frac{Id\vec{l}\times\hat{r}}{r^2}$ $\quad$ $\mu = IA$

$|\text{emf}| = \left|\int_{\text{closed path}} \vec{E}\cdot d\vec{l}\right| = \left|\frac{d\Phi_{magnetic}}{dt}\right|$ $\quad$ $\Phi_{magnetic} = \int_{\text{bounded surface}} \vec{B}\cdot\hat{n}dA$ $\quad$ $\text{P.E.} = -\vec{\mu}\cdot\vec{B}$

$\vec{F} = q\vec{E} + q\vec{v}\times\vec{B}$ $\quad$ $d\vec{F} = Id\vec{l}\times\vec{B}$

$\vec{E}_{radiative} = \frac{1}{4\pi\varepsilon_0}\frac{-q\vec{a}_\perp}{c^2 r}$, propagates in direction of $\vec{E}\times\vec{B}$

## MECHANICS

$\vec{F} = m\vec{a}$ $\quad$ $a_{circular} = \frac{v^2}{r}$

$$E_{\text{point or sphere}} = \frac{1}{4\pi\varepsilon_0}\frac{Q}{d^2}$$

$$E_{\text{dipole}} \approx \frac{1}{4\pi\varepsilon_0}\left[\frac{2qs}{d^3}\right]$$ at a distance d from the dipole midpoint, along the axis of the dipole

$$E_{\text{dipole}} \approx \frac{1}{4\pi\varepsilon_0}\left[\frac{qs}{d^3}\right]$$ at a distance d from the dipole midpoint, perpendicular to axis of the dipole

dipole moment $p = qs = \alpha E$, where $\alpha$ is the "polarizability"

$$E_{\text{rod}} = \frac{Q}{4\pi\varepsilon_0}\left[\frac{1}{x\sqrt{x^2+(L/2)^2}}\right]$$, perp. to rod

$$E_{\text{rod}} \approx \frac{1}{4\pi\varepsilon_0}\frac{2Q/L}{x}$$, if $x << L$

$$E_{\text{ring}} = \frac{1}{4\pi\varepsilon_0}\frac{qx}{(x^2+r^2)^{3/2}}$$, along axis of ring

$$E_{\text{disk}} = \frac{Q/A}{2\varepsilon_0}\left[1-\frac{x}{\left(x^2+R^2\right)^{1/2}}\right]$$

$$E_{\text{disk}} \approx \frac{Q/A}{2\varepsilon_0}$$, if $x << R$

$$E_{\text{capacitor}} \approx \frac{Q/A}{\varepsilon_0}$$ (+Q disk and -Q disk)

$$E_{\text{fringe}} \approx \frac{Q/A}{\varepsilon_0}\left(\frac{s}{2R}\right)$$ just outside capacitor

$$B_{\text{wire}} \approx \frac{\mu_0}{4\pi}\frac{2I}{x}$$ a distance x from a long straight wire

$$B \approx \frac{\mu_0}{4\pi}\frac{2\mu}{x^3}$$, with $\mu = IA$, along the axis a distance x, far from a current loop

for circular loop of radius a, current I, a distance x from the center, along the axis

$$B_{\text{solenoid}} \approx \frac{\mu_0 NI}{L}$$ inside a long solenoid with N turns (radius of coil << length L)

$$\int\frac{dz}{\left(z^2+a^2\right)^{1/2}} = \ln\left(z+\sqrt{z^2+a^2}\right)$$

$$\int\frac{dz}{z\left(z^2+a^2\right)^{1/2}} = -\frac{1}{a}\ln\left(\frac{z+\sqrt{z^2+a^2}}{z}\right)$$

$$\int\frac{zdz}{\left(z^2+a^2\right)^{1/2}} = \sqrt{z^2+a^2}$$

$$\int\frac{dz}{\left(z^2+a^2\right)^{3/2}} = \frac{z}{a^2\sqrt{z^2+a^2}}$$

$$\int\frac{zdz}{\left(z^2+a^2\right)^{3/2}} = \frac{-1}{\sqrt{z^2+a^2}}$$

$$\int\frac{z^2dz}{\left(z^2+a^2\right)^{3/2}} = \frac{-z}{\sqrt{z^2+a^2}} + \ln\left(z+\sqrt{z^2+a^2}\right)$$

# CHAPTER 7

# SAMPLE QUIZZES & EXAMS (SEMESTER B)

Here we present quizzes, exams, and the final exam for an earlier semester (semester "B"), one in which the weekly quizzes came after the homework problems had been turned in. The quizzes were more difficult than in semester "A" and a bit more like homework and exam problems.

Quizzes were nominally 15 to 20 minutes. Exams were nominally an hour long, but those students who needed more time were given an additional half-hour or hour. The final exam was nominally three hours long, but those students who needed more time were given an additional hour or two.


Quiz for Chapter 2 92

Quiz for Chapter 3 93

Another quiz for Chapter 3 93

Quiz for Chapter 5 94

Quiz for Chapter 8 95

Quiz for Chapter 11 95

Quiz for Chapter 12 96

Quiz for Chapter 13 96

Another quiz for Chapter 13 97

Quiz for Chapter 14 97

Exam 1 covering Chapters 1 through 4 98

Exam 2 covering Chapters 5 through 9 102

Exam 3 covering Chapters 10 through 13 106

Final exam covering Chapters 1 through 14 109

## Quiz for Chapter 2

A proton interacts with a distant electron. How does the *net force on the electron* change if a piece of paper is inserted between the proton and the electron? Does the net force on the electron get bigger, smaller, or stay the same? **Explain *briefly* but carefully, in terms of basic principles, and use diagrams to help in the explanation.**

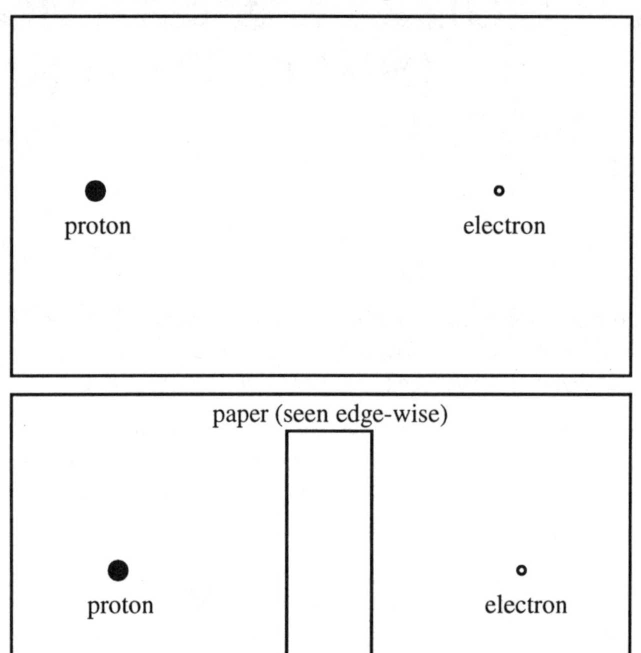

## Quiz for Chapter 3

Place a *helium nucleus* and a *proton* somewhere in the box in such a way that the electric field due to these charges is zero at the location marked ×. (A helium nucleus contains two protons and two neutrons.) **Explain *briefly* but carefully, and use diagrams to help in the explanation.**

Place a *helium nucleus* and an *electron* somewhere in the box in such a way that the electric field due to these charges is zero at the location marked ×. **Explain *briefly* but carefully, and use diagrams to help in the explanation.**

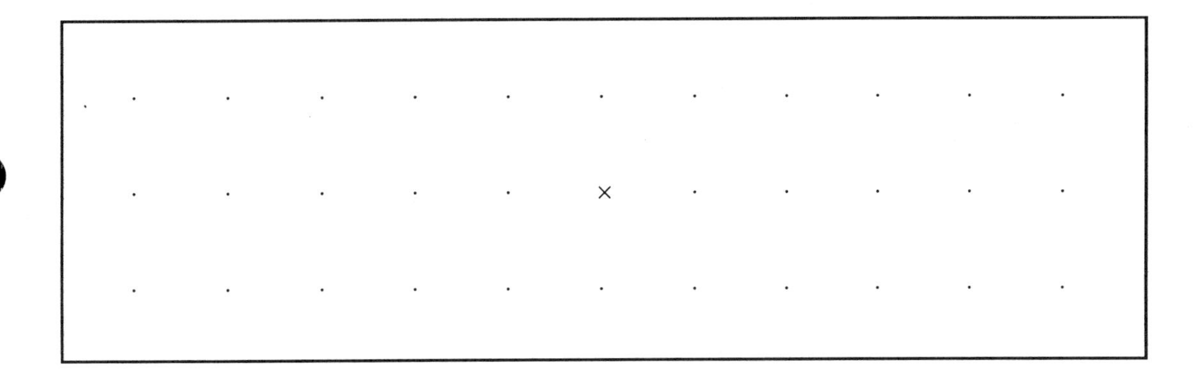

## Another quiz for Chapter 3

You are the captain of a spaceship. You need to measure the electric field at a specified location P in space outside your ship. You send a crew member outside with a meter stick, a stopwatch, and a small ball of known mass M and net charge +Q (held by insulating strings while being carried). Explain to the crew member what observations of the ball to make for you, and explain how you will use those observations to determine the magnitude and direction of the electric field at location P.

a) Instructions to crew member:

b) How you will analyze the data:

## Quiz for Chapter 5

At all locations indicated on the diagrams below, *draw the direction the compass needle will point*, and *enter the approximate magnitude of the compass deflection, in degrees*, in the box, like this:

Note that the deflection is given at one location in circuit #1. If you do not have enough information to give a number, then indicate whether it will be greater than, less than, or equal to 9 degrees. *In each case, the wire is laid on top of the compass. Briefly* explain your reasoning about the magnitude of the deflection in each case.

When it is far from the circuit, the compass points this way: 

1)

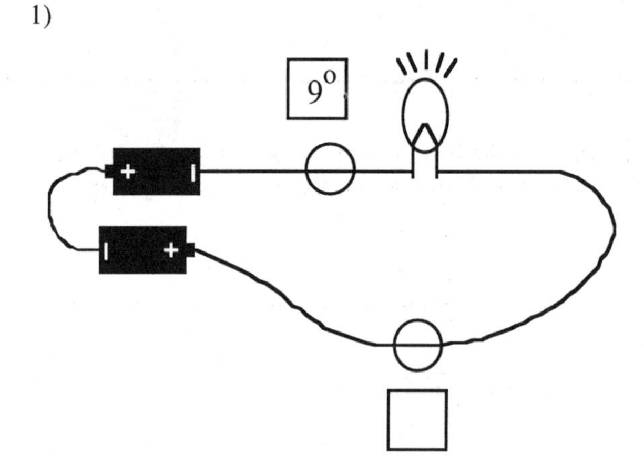


2)

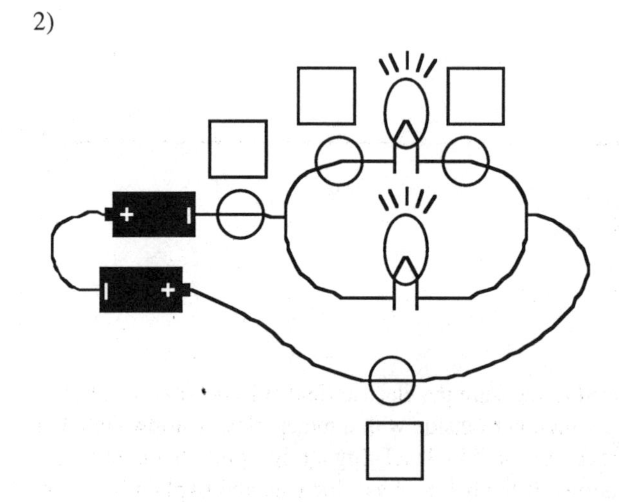

## Quiz for Chapter 8

A voltmeter measured a potential difference $V_2 - V_1 = +300$ volts between two plates of a capacitor with very large plates and a gap distance of 0.2 millimeters.

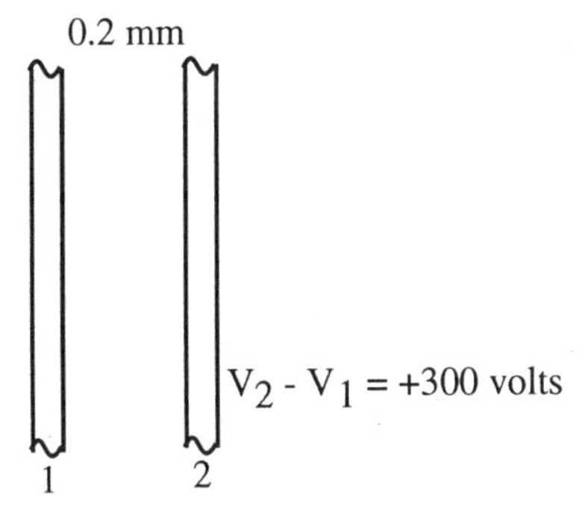


(a) Draw the electric field at a location in the gap, and draw the charges on the capacitor plates.

(b) Calculate the magnitude of the electric field. Explain briefly.

(c) An electron is released in the gap very near the negative plate. How much kinetic energy does it have when it crashes into the positive plate?

---

$\frac{1}{4\pi\varepsilon_0} = 9 \times 10^9 \, N - m^2 / C^2$ $\quad e = 1.6 \times 10^{-19}$ C $\quad c = 3 \times 10^8$ m/s

## Quiz for Chapter 11

Here are a bar magnet and a compass:

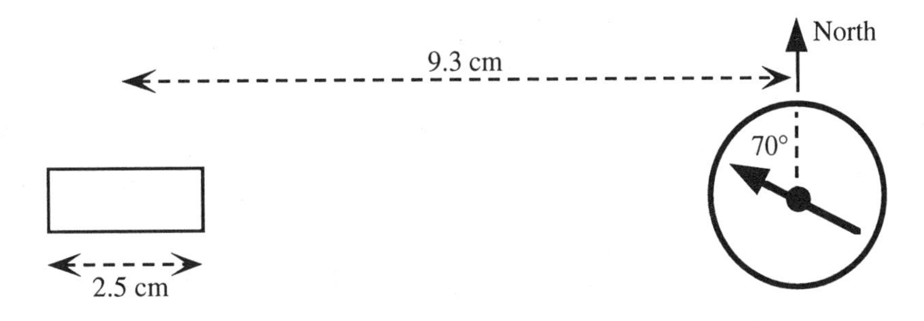


(a, 5 pts) Draw the magnetic field *due to the bar magnet* at the center of the compass.

(b, 5 pts) Label the N and S poles of the bar magnet.

(c, 10 pts) Calculate the magnitude of the magnetic field *due to the bar magnet* at the center of the compass. Explain your calculation.

---

horizontal component of $B_{earth} = 2 \times 10^{-5}$ tesla $\quad \frac{\mu_0}{4\pi} = 10^{-7}$ tesla – m / ampere

## Quiz for Chapter 12

An electron travels in a straight line at constant speed $v = 4\times10^7$ m/sec in the x-direction through a region of uniform electric and magnetic fields. The electric field is up, along the y-axis, with magnitude E = 6000 volts/meter.

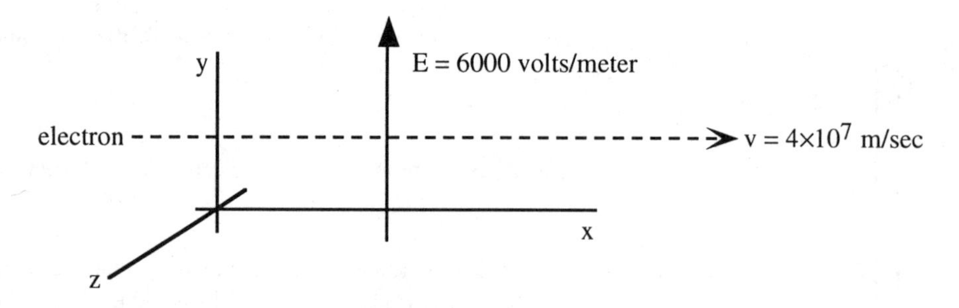


(a, 5 pts) What is the direction of the magnetic field?

(b, 10 pts) What is the magnitude B of the magnetic field?

(c, 5 pts) Explain how you could add an additional magnetic field and not affect the motion of the electron.

## Quiz for Chapter 13

The north pole of a bar magnet points toward a thin circular coil of wire containing 40 turns. The magnet is moved away from the coil, so that the flux inside the coil decreases by 0.3 tesla-$m^2$ in 0.2 seconds.

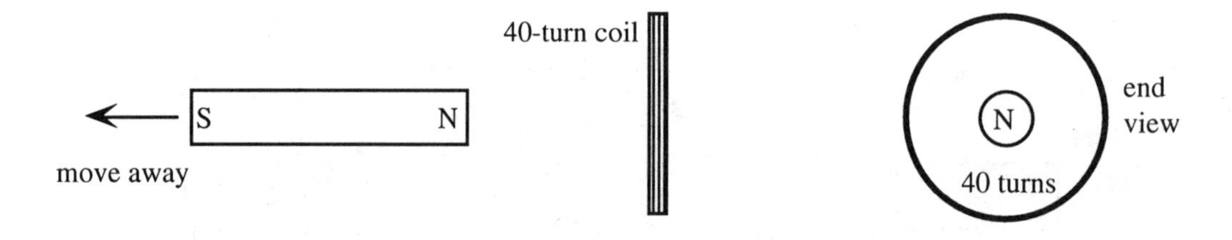


What is the average emf induced in the coil during this time interval? Viewed from the right side (opposite the bar magnet), does the induced current run clockwise or counterclockwise? Explain briefly, including appropriate diagrams.

## Another quiz for Chapter 13

A metal rod of length L is dragged horizontally at constant speed v on frictionless *conducting* rails through a region of uniform magnetic field of magnitude B, pointing into the page.

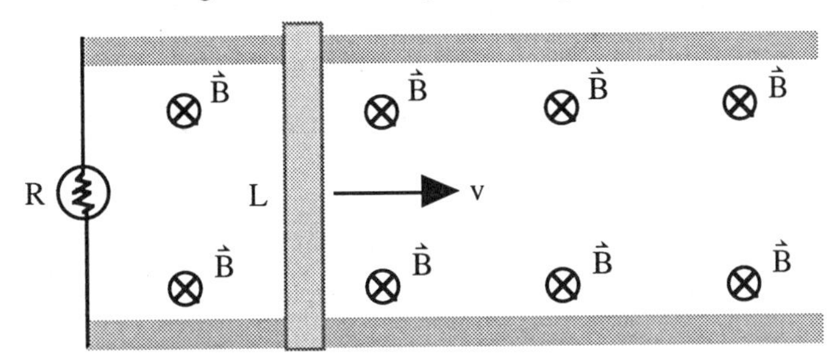


1) Show the direction of the induced current I. Explain briefly.

2) What is the magnitude of the current I? Explain briefly.

3) What is the magnitude and direction of the force $\vec{F}$ you have to apply to keep the rod moving at a constant speed v? Explain briefly.

## Quiz for Chapter 14

If the electric field inside a capacitor exceeds $3\times10^6$ V/m (approximately), the few free electrons in the air are accelerated enough to trigger an avalanche and make a spark. In the spark shown in the diagram, electrons are accelerated downward and positive ions are accelerated upward. Qualitatively, explain the directions and relative magnitudes of the radiative electric fields off to the right of the capacitor, due to the motions of the electrons and ions:

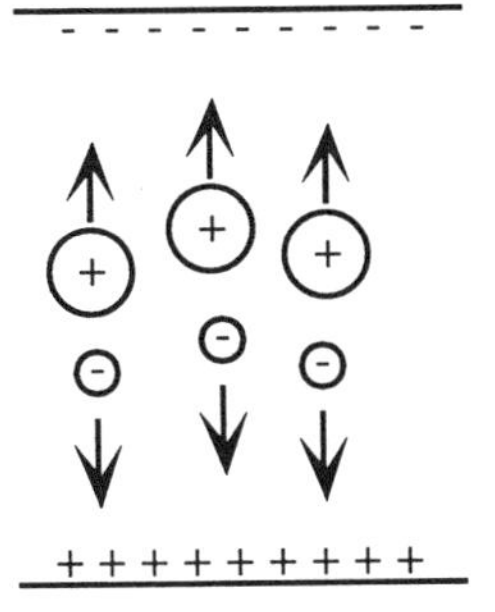

$\vec{E}$ ?

If you are 10 meters away and to the right of the capacitor, how long after the initiation of the spark could you first detect a magnetic field? (Before the spark occurs there is no magnetic field anywhere in this region.)

What is the direction of the radiative magnetic field?

---

$\vec{E}_{radiative} = \frac{1}{4\pi\varepsilon_0}\frac{-q\vec{a}_\perp}{c^2 r}$ $\qquad$ $c = 3\times10^8$ m/s $\qquad$ direction of propagation is $\vec{E}\times\vec{B}$

## Exam 1 covering Chapters 1 through 4

- **Read all problems carefully before attempting to solve them.**
- **Correct answers without adequate explanation *will be counted wrong*.**
- **Incorrect explanations mixed in with correct explanations *will be counted wrong*.**
- **Make explanations complete but brief. Do not write a lot of prose.**
- **Include diagrams where needed.**

***There is a tear-off formula sheet on the back.***

**Problem 1 (35 pts):** A metal sphere of radius 0.25 m carries an initial charge $Q_1 = +1.4\times10^{-5}$ coulomb.

**(a, 10 pts):** Prove that this amount of charge is insufficient to ionize the surrounding air.

**(b, 15 pts):** Your body is initially neutral, and the soles of your shoes are made of insulating material. You move your finger close to the sphere but without touching the sphere. When your finger gets very close to the sphere you see a spark in the air between your finger and the sphere. Explain why a spark appears, despite $Q_1$ being insufficient to ionize the air before you brought your finger near. *Remember to include appropriate diagrams!*

**(c, 5 pts):** After seeing the spark you move your finger away from the sphere. The charge on the sphere is now $Q_2$.

Is $Q_2$ less than, equal to, or greater than $Q_1$? Explain briefly.

**(d, 5 pts):** In terms of the given quantities, what is *your* net charge now? Explain briefly.

**Problem 2 (30 pts):** A small, thin, hollow spherical glass shell of radius R carries a uniformly distributed positive charge +Q. Below it is a horizontal permanent dipole with charges +q and -q separated by a distance s (s is shown greatly enlarged for clarity). The dipole is fixed in position and is not free to rotate The distance from the center of the glass shell to the center of the dipole is L.

**(a, 15 pts):** Calculate the **magnitude** and **direction** of the electric field at the center of the glass shell, and explain briefly, *including showing the electric field on the diagram*. Your results *must not contain* any symbols other than the given quantities R, Q, q, s, and L (and fundamental constants), unless you define intermediate results in terms of the given quantities. What simplifying assumption do you have to make?

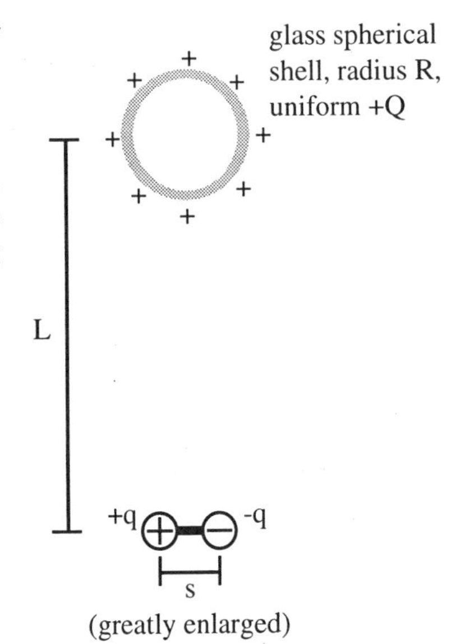


**(b, 15 pts):** If the upper sphere were a solid metal ball with a charge +Q, what would be the **magnitude** and **direction** of the electric field at its center? (A slice through the solid ball is shown.) Explain briefly. Show the distribution of charges everywhere, and at the center of the ball *accurately* draw and label the electric field $\vec{E}_{ball}$ due to the ball charges and the electric field of the dipole $\vec{E}_{dipole}$.

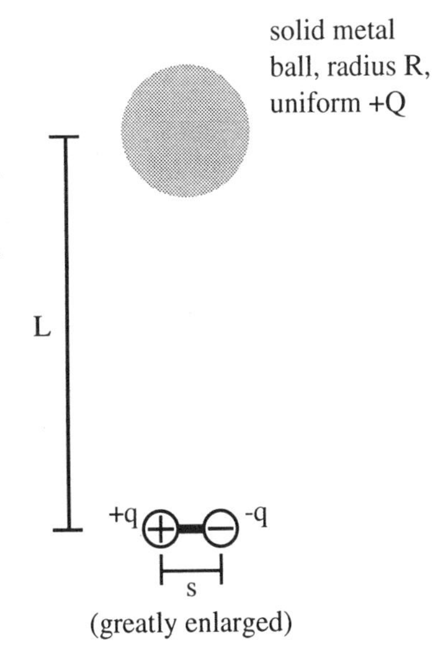

**Problem 3 (35 pts):** A long thin plastic rod of length L is horizontal and centered at the origin. It is uniformly charged negatively, with a total charge -Q. Use the standard procedure to determine the magnitude and direction of the electric field at a location on the (downward-pointing) y-axis, a distance y from the origin. You do *not* have to evaluate the integral that you obtain: you can get that result from the formula page. But you must carry out all other aspects of the procedure, including showing that the result is reasonable. Be sure to draw appropriate diagrams!

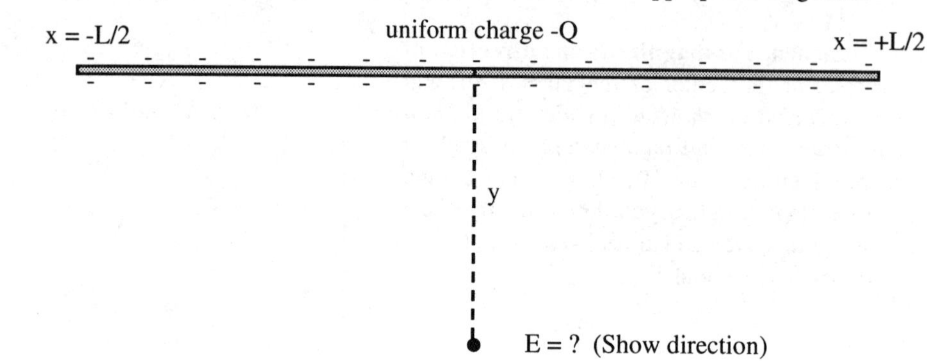

**Formula sheet for Exam 1**

$\frac{1}{4\pi\varepsilon_0} = 9\times10^9$ N-m²/C² $\quad$ e = $1.6\times10^{-19}$ coulomb

$\varepsilon_0 = 9\times10^{-12}\ C^2/N\text{-}m^2$ $\quad$ g = 9.8 N/kg

$m_{\text{hydrogen atom}} = \frac{(1\times10^{-3}\text{ kilogram/mole})}{(6\times10^{23}\text{ atoms/mole})}$ $\quad$ $m_{\text{electron}} = 9\times10^{-31}$ kg

$E_{\text{point or sphere}} = \frac{1}{4\pi\varepsilon_0}\frac{Q}{d^2}$ $\quad$ $E_{\text{breakdown in air}} \approx 3\times10^6$ N/C

$E_{\text{dipole}} \approx \frac{1}{4\pi\varepsilon_0}\left[\frac{2sq}{d^3}\right]$ at a location a distance d from the dipole midpoint, along the axis

$E_{\text{dipole}} \approx \frac{1}{4\pi\varepsilon_0}\left[\frac{sq}{d^3}\right]$ at a location a distance d from the dipole midpoint, perpendicular to the axis

$E_{\text{rod}} = \frac{Q}{4\pi\varepsilon_0}\left[\frac{1}{x\sqrt{x^2+(L/2)^2}}\right]$, perp. to rod $\quad$ $E_{\text{rod}} \approx \frac{1}{4\pi\varepsilon_0}\frac{2Q/L}{x}$, if x << L

$E_{\text{ring}} = \frac{1}{4\pi\varepsilon_0}\frac{qx}{(x^2+r^2)^{3/2}}$ $\quad$ area of circle = $\pi r^2$

$E_{\text{disk}} = \frac{Q/A}{2\varepsilon_0}\left[1-\frac{x}{\left(x^2+R^2\right)^{1/2}}\right]$ $\quad$ $E_{\text{disk}} \approx \frac{Q/A}{2\varepsilon_0}$, if x << R

$E_{\text{capacitor}} \approx \frac{Q/A}{\varepsilon_0}$ (+Q disk and -Q disk) $\quad$ $E_{\text{fringe}} \approx \frac{Q/A}{\varepsilon_0}\left(\frac{s}{2R}\right)$ just outside the capacitor

## Exam 2 covering Chapters 5 through 9

- **Read all problems carefully before attempting to solve them.**
- **Correct answers without adequate explanation** ***will be counted wrong.***
- **Incorrect explanations mixed in with correct explanations** ***will be counted wrong.***
- **Make explanations complete but brief. Do not write a lot of prose.**
- **Include diagrams where needed.**

| *There is a tear-off formula sheet on the back.* |
|---|
| *Use only these formulas! Derive anything else.* |

**Problem 1 (50 pts).** A 40-cm-long high-resistance wire with rectangular cross section (7 mm by 3 mm) is connected to a 12-volt battery through an ammeter. The resistance of the wire is 50 ohms. The resistance of the ammeter and the internal resistance of the battery can be considered to be negligibly small compared to the resistance of the wire.

Leads to a voltmeter are connected as shown, with the "+" lead connected to the outer edge of the wire, at the bottom (location "b"), and the "-" lead connected to the inner edge of the wire, at the top (location "d"). The distance along the wire between voltmeter connections is 5 cm.

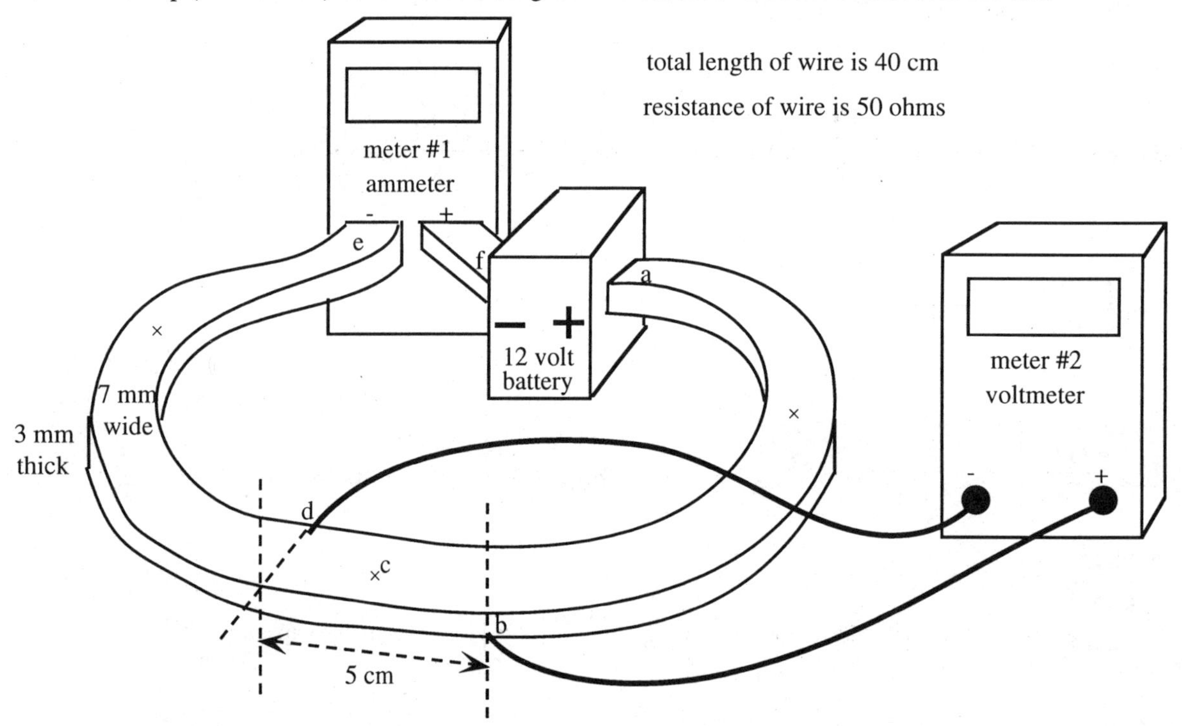


**(a, 5 pts)** On the diagram, show the approximate distribution of charge.

**(b, 5 pts)** On the diagram, draw the electric field *inside* the wire at the 3 positions marked ×.

**(c, 5 pts)** What is the magnitude of the electric field at location c?

**(d, 5 pts)** What does the voltmeter read, both magnitude and sign?

**(e, 5 pts)** What does the ammeter read, both magnitude and sign?

**(f, 5 pts)** In a 60-second period, how many electrons are released from the "-" end of the battery?

**(g, 5 pts)** There are $1.5\times10^{26}$ free electrons per cubic meter in the wire. What is the drift speed v of the electrons in the wire?

**(h, 5 pts)** What is the mobility u of the material that the wire is made of?

**(i, 5 pts)** Switch meter #1 from being an ammeter to being a voltmeter. Now what do the two meters read?

**(j, 5 pts)** The 12-volt battery is removed from the circuit and both the ammeter and voltmeter are connected in parallel to the battery. The voltmeter reads 1.8 volts, and the ammeter reads 20.4 amperes. What is the internal resistance of the battery?

**Problem 2 (35 pts).**

A thin spherical shell made of plastic carries a uniformly distributed negative charge $-Q_1$. Two large disks made of glass carry uniformly distributed positive and negative charges $+Q_2$ and $-Q_2$. The radius $R_1$ of the plastic spherical shell is very small compared to the radius $R_2$ of the glass disks. The distance from the center of the spherical shell to the positive disk is d, and d is much smaller than $R_2$.

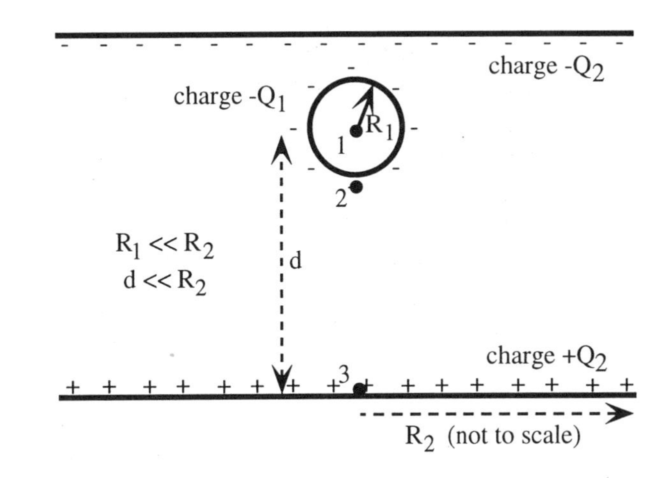


**(a, 10 pts)** Find the potential difference $V_2 - V_1$ in terms of the given quantities ($Q_1$, $Q_2$, $R_1$, $R_2$, and d). Location 1 is at the center of the plastic sphere, and location 2 is just outside the sphere.

**(b, 15 pts)** Find the potential difference $V_3 - V_2$. Location 2 is just below the sphere, and location 3 is right beside the positive glass disk. Do a *complete* calculation, starting only from formulas on the formula sheet.

**(c, 10 pts)** Suppose the plastic shell is replaced by a solid metal sphere with radius $R_1$ carrying charge $-Q_1$. State whether the absolute magnitudes of the potential differences would be greater than, less than, or the same as they were with the plastic shell in place. Explain briefly, including an appropriate diagram.

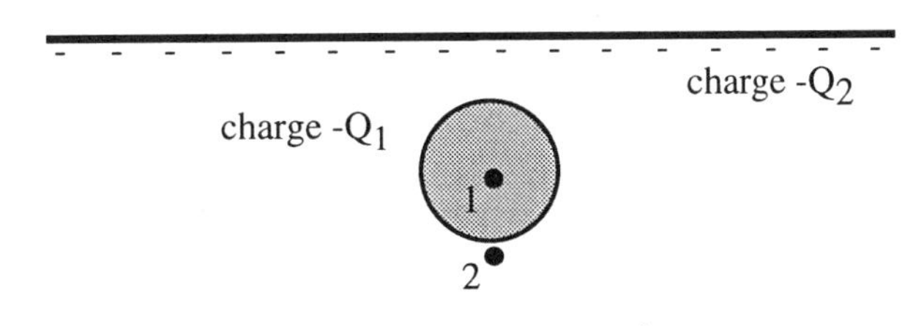


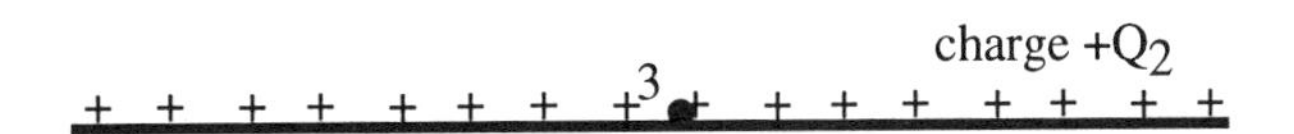


$|V_2 - V_1|$ would be greater than, less than, or the same?

$|V_3 - V_2|$ would be greater than, less than, or the same?

**Problem 3 (15 pts).** Here is a circuit composed of two batteries, two identical long bulbs, and an initially uncharged capacitor. Three compasses are placed on the desktop with the wires running on top of the compasses. All the compasses point north before the gap is closed. Then the gap is closed, and the compass on the left immediately shows about a 20 degree deflection.

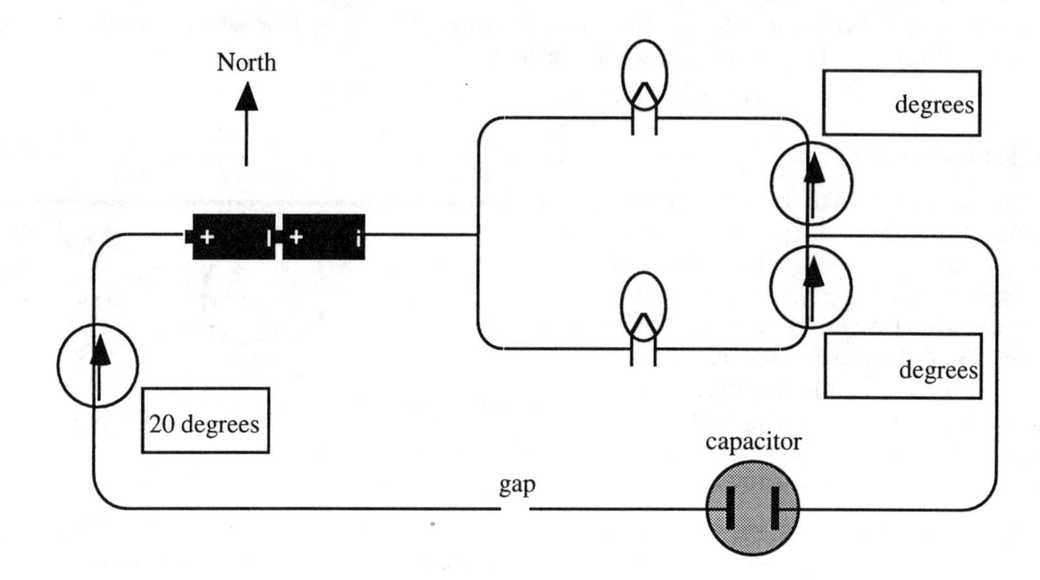


**(a, 8 pts)** Draw the compass needle positions on all three compasses at the instant when the left-most compass shows a 20-degree deflection, and write the approximate angle of deflection beside the other two compasses.

**(b, 7 pts)** When this experiment was performed with one of the bulbs removed from its socket, the single bulb glowed for 15 seconds. Predict how long the two bulbs would glow when both are in their sockets (starting with the capacitor discharged).

**Formula sheet for Exam 2**

$\frac{1}{4\pi\varepsilon_0} = 9\times10^9\ \text{N-m}^2/\text{C}^2$

$e = 1.6\times10^{-19}$ coulomb

$\varepsilon_0 = 9\times10^{-12}\ \text{C}^2/\text{N-m}^2$

$g = 9.8$ N/kg

$m_{\text{hydrogen atom}} = \frac{(1\times10^{-3}\ \text{kilogram/mole})}{(6\times10^{23}\ \text{atoms/mole})}$

$m_{\text{electron}} = 9\times10^{-31}$ kg

$E_{\text{point charge}} = \frac{1}{4\pi\varepsilon_0}\frac{Q}{r^2}$

$E_{\text{breakdown in air}} \approx 3\times10^6$ N/C

Very near flat sheet of charge: $E \approx \frac{Q/A}{2\varepsilon_0}$

$\Delta V = \frac{\Delta(PE)}{q} = -\int_i^f \vec{E}\bullet d\vec{l}$

Two-disk capacitor, at center (for d << R):

inside: $E \approx \frac{Q/A}{\varepsilon_0}$

just outside: $E \approx \frac{Q/A}{\varepsilon_0}\frac{s}{2R}$

$E = -\frac{\Delta V}{\Delta L}$

$Q = C|\Delta V|$

Dipole with charge q and -q separated by s along x:

$E_x = \frac{1}{4\pi\varepsilon_0}\frac{2sq}{x^3}$ along x-axis

$E_x = -\frac{1}{4\pi\varepsilon_0}\frac{sq}{y^3}$ along y-axis

$\Delta V_{\text{round trip}} = 0$

$\Sigma(I) = 0$

power $= I\times\Delta V$

#/sec = nAv; v = uE

$R = \frac{L}{\sigma A}$

$J = qnv = \sigma E$; $I = JA$

$I = \frac{\Delta V}{R}$

$\Delta V_{\text{battery}} = \text{emf} - r_{\text{internal}}I$

$E \rightarrow E/K$ inside insulator

$Q = Q_{\text{final}}\left[1 - e^{-\frac{t}{RC}}\right]$ (charging)

$Q = Q_{\text{initial}}e^{-\frac{t}{RC}}$ (discharging)

circumference of circle $= 2\pi r$    area of circle $= \pi r^2$    volume of sphere $= \frac{4\pi}{3}r^3$

## Exam 3 covering Chapters 10 through 13

- **Read all problems carefully before attempting to solve them.**
- **Correct answers without adequate explanation** ***will be counted wrong.***
- **Incorrect explanations mixed in with correct explanations** ***will be counted wrong.***
- **Make explanations complete but brief. Do not write a lot of prose.**
- **Include diagrams where needed.**

***There is a tear-off formula sheet on the back.***

**Problem 1 (30 pts).**

**(a, 15 pts)** An electron is moving with a speed v in the plane of the page, and there is a uniform magnetic field B into the page throughout this region. Draw the trajectory of the electron, and explain qualitatively.

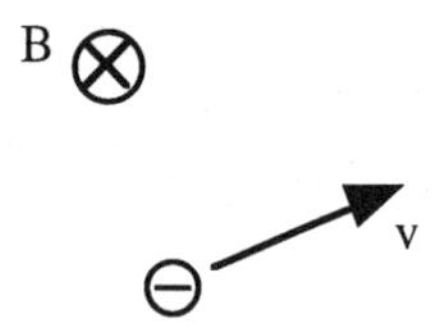


**(b, 15 pts)** A bar magnet is mounted on a pivot as shown and released in the presence of a magnetic field $\vec{B}$ made by coils that are not shown. Does the magnet initially turn clockwise or counter-clockwise? Explain qualitatively in detail why it turns, and why it turns in the predicted direction. *Do not use any formulas that are not on the formula page.*

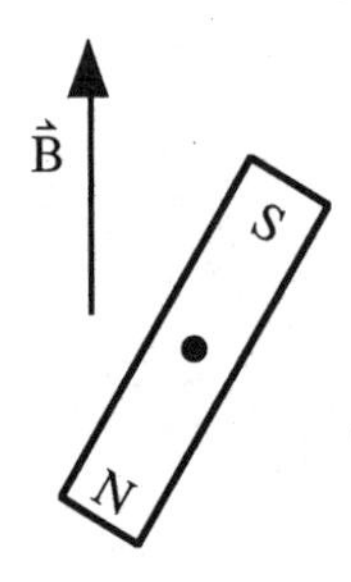

**Problem 2 (35 pts).** At a particular instant, a proton and electron are a distance L apart, with the proton heading in the +y direction with speed $v_p$ and the electron headed in the -x direction with speed $v_e$. Calculate the x- and y-components of the forces on the two particles. Be clear in your explanation, including clear and understandable diagrams, and use the given symbols in your answers.

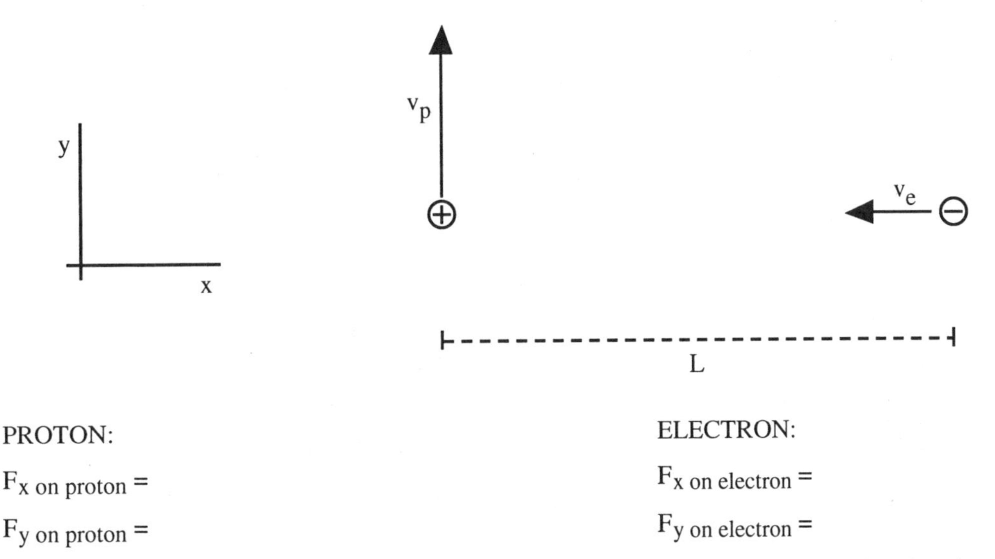


PROTON:

$F_{x \text{ on proton}}$ =

$F_{y \text{ on proton}}$ =

ELECTRON:

$F_{x \text{ on electron}}$ =

$F_{y \text{ on electron}}$ =

(Newton's Third Law does not apply here: some momentum is carried by the electric and magnetic fields.)

**Problem 3 (35 pts).** A thin circular coil of radius $r_1$ with $N_1$ turns carries a current $I_1 = a + bt + ct^2$, where t is the time and a, b, and c are positive constants. A second thin circular coil of radius $r_2$ with $N_2$ turns is located a long distance x along the axis of the first coil. The second coil is connected to an oscilloscope, which has very high resistance.

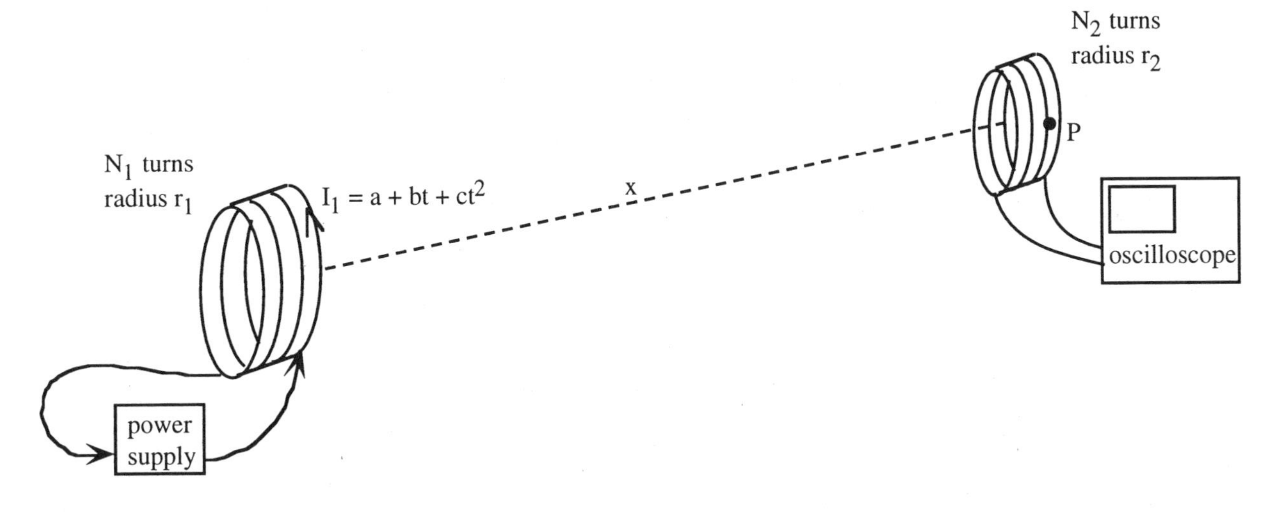


**(a, 25 pts)** As a function of time t, calculate the magnitude of the voltage that is displayed on the oscilloscope. Explain your work carefully, but you do not have to worry about signs.

**(b, 5 pts)** At location P on the drawing (on the right side of the second coil), draw a vector representing the *non-Coulomb electric* field.

**(c, 5 pts)** Calculate the magnitude of this *non-Coulomb electric* field.

**Formula sheet for Exam 3**

$\frac{1}{4\pi\varepsilon_0} = 9\times10^{9}\ \text{N-m}^2/\text{C}^2$

$e = 1.6\times10^{-19}$ coulomb

$\varepsilon_0 = 9\times10^{-12}\ \text{C}^2/\text{N-m}^2$

$g = 9.8$ N/kg

$\frac{\mu_0}{4\pi} = 10^{-7}\ \frac{\text{tesla-m-sec}}{\text{C}}$

$m_{electron} = 9\times10^{-31}$ kg

circle circumference = $2\pi r$

circle area = $\pi r^2$

$$\oint_{\text{closed surface}} \vec{E}\bullet\hat{n}dA = \frac{\sum q_{inside}}{\varepsilon_0}$$

$$\oint_{\text{closed surface}} \vec{B}\bullet\hat{n}dA = 0 \text{ (Gauss's Law for magnetism)}$$

$$\vec{F} = q\vec{E} + q\vec{v}\times\vec{B}$$

$$d\vec{F} = Id\vec{l}\times\vec{B}$$

$$\vec{B} = \frac{\mu_0}{4\pi}\frac{q\vec{v}\times\hat{r}}{r^2}$$

$$d\vec{B} = \frac{\mu_0}{4\pi}\frac{Id\vec{l}\times\hat{r}}{r^2}$$

$$\text{emf}_{\text{along bounding path}} = \oint_{\text{along bounding path}} \vec{E}\bullet d\vec{l} = -N\frac{d}{dt}\left(\int_{\text{open surface}} \vec{B}\bullet\hat{n}dA\right)$$

$B_{wire} = \frac{\mu_0}{4\pi}\frac{2aI}{x\sqrt{x^2+a^2}}$ for straight wire, length 2a, current I, a distance x from the center of wire

$B \approx \frac{\mu_0}{4\pi}\frac{2IA}{x^3}$ along the axis a distance x, far from a current loop

$B_{solenoid} \approx \frac{\mu_0 NI}{L}$ inside a long solenoid with N turns (radius of coil << length L)

## Final exam covering Chapters 1 through 14

***The last page is a tear-off list of fundamental equations and constants.***

**Show all your work on the exam pages. You must provide supporting calculations and explanations. Correct answers without such support are not acceptable.**

On some problems you must use only formulas from the formula sheet.
On other problem you may use information from your notes,
but you must explain what these formulas are.

**Prob. 1 (70 pts):**

**(a, 10 pts)** Estimate the magnitude of electric field required to pull an outer electron out of an atom. (All atoms have a diameter very roughly equal to $10^{-10}$ meters.)

**(b, 10 pts)** Explain why can you get a spark in air with a much smaller electric field than the magnitude you just calculated.

**(c, 10 pts)** With one 1.5-volt battery and a particular light bulb, 0.25 ampere runs through the bulb. With two batteries in series with the same bulb, the current is only 0.35 ampere, even though the internal resistance of the batteries is negligible compared to the resistance of the bulb. Explain briefly.

**(d, 10 pts)** Sketch the approximate surface-charge distribution on this circuit:

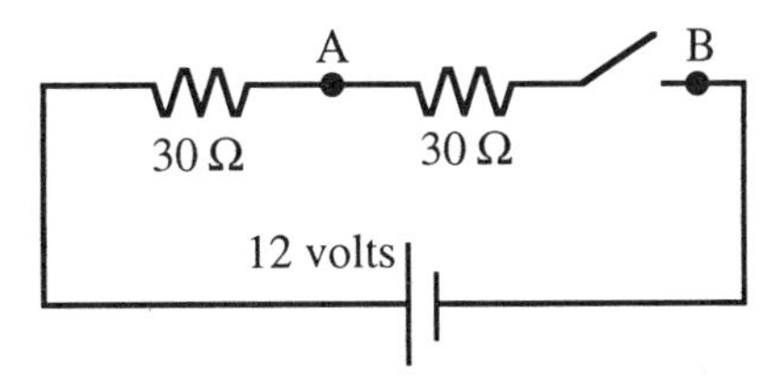


**(e, 10 pts)** What is the potential difference between locations A and B on the circuit shown in part (d)? Explain your reasoning.

**(f, 10 pts)** What is the direction of the magnetic field at location P due to the conventional current in two long wires that meet at a corner? Explain briefly.

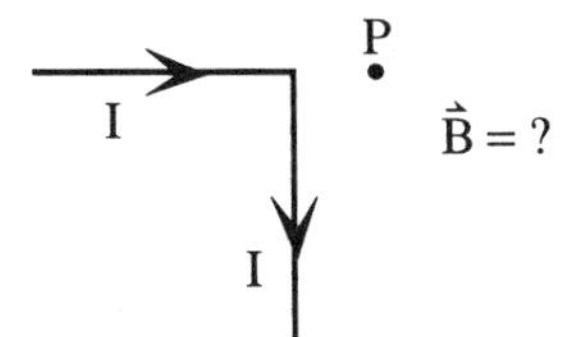

**(g, 10 pts)** The short-wave radio transmitter demonstrated in class has a horizontal antenna consisting of two straight wires connected to a high-frequency AC voltage. Consider three different positions of the receiving antenna, which consists of two straight wires connected through a flashlight bulb:

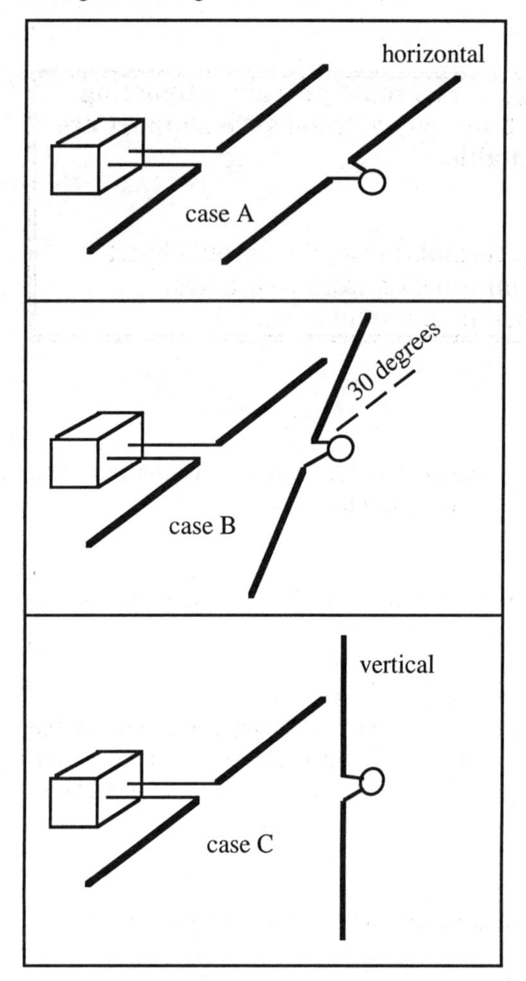


Describe the brightness of the bulb in the three cases (case A, B, and C), including a ranking of the relative brightness (which is brightest, which is dimmest, whether any cases are the same brightness):

Explain briefly why the brightnesses are what they are:

**Prob. 2 (45 pts):**

**(a, 10 pts)** A glass sphere carrying a uniformly-distributed charge of +Q is surrounded by an initially neutral spherical *plastic* shell. Qualitatively, indicate the polarization of the plastic.

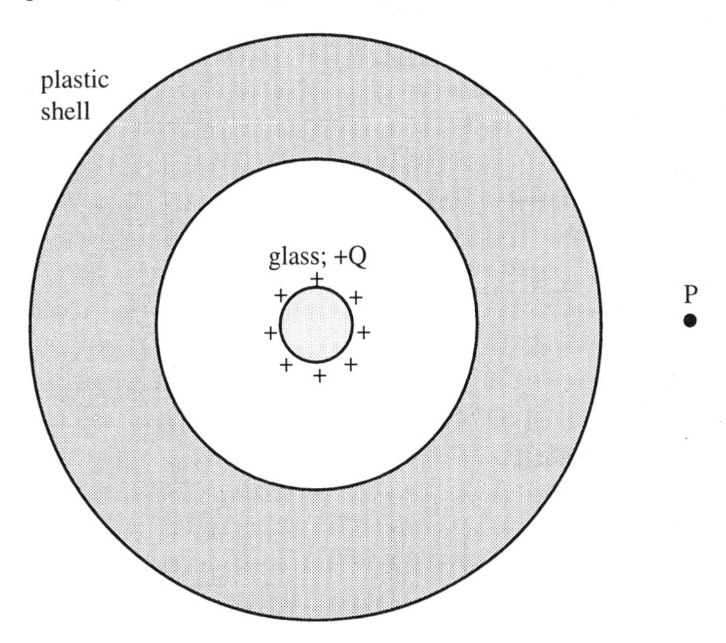


**(b, 5 pts)** Qualitatively, indicate the polarization of the inner glass sphere. Explain briefly.

**(c, 5 pts)** Is the electric field at location P outside the plastic shell larger, smaller, or the same as it would be if the plastic weren't there? Explain briefly.

**(d, 10 pts)** Now suppose that the glass sphere carrying a uniform charge of +Q is surrounded by an initially neutral *metal* shell. Qualitatively, indicate the polarization of the metal.

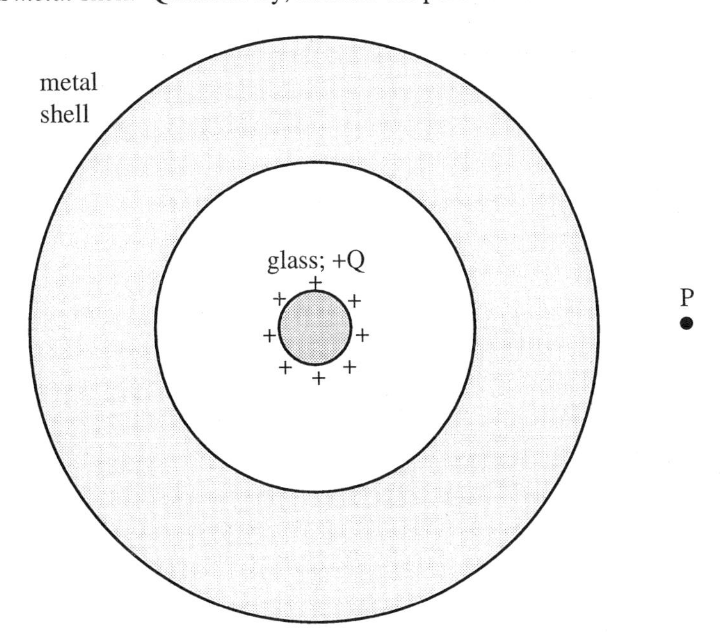


**(e, 10 pts)** Now be *quantitative* about the polarization of the metal sphere and *prove* your assertions.

**(f, 5 pts)** Is the electric field at location P outside the metal shell larger, smaller, or the same as it would be if the metal shell weren't there? Explain briefly.

**Prob. 3 (40 pts):** A thin plastic rod is bent into a half-circle of radius R and charged uniformly with a negative charge -Q. Find the magnitude and direction of the electric field at the center of the half-circle. Explain all your steps carefully and clearly. Remember that arc length is equal to radius times angle (as long as the angle is measured in radians).

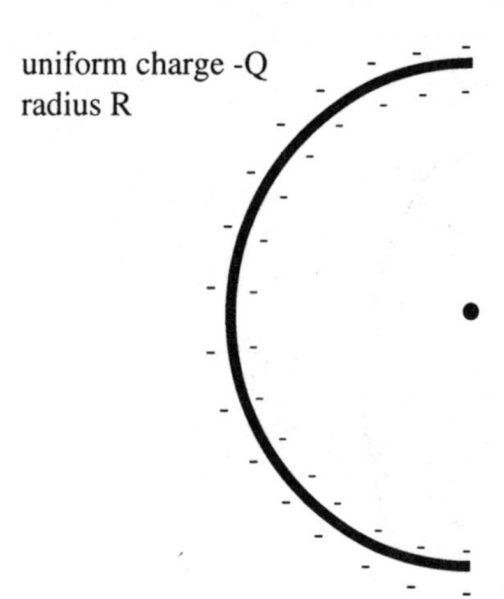


**Prob. 4 (35 pts):** A short distance x away from the center of a very long uniformly charged rod of length L (with x << L), the electric field is radial and has the magnitude $E \approx \frac{1}{4\pi\varepsilon_0}\frac{2(Q/L)}{x}$.

Calculate the potential difference $V_A - V_C$, and explain your work carefully.

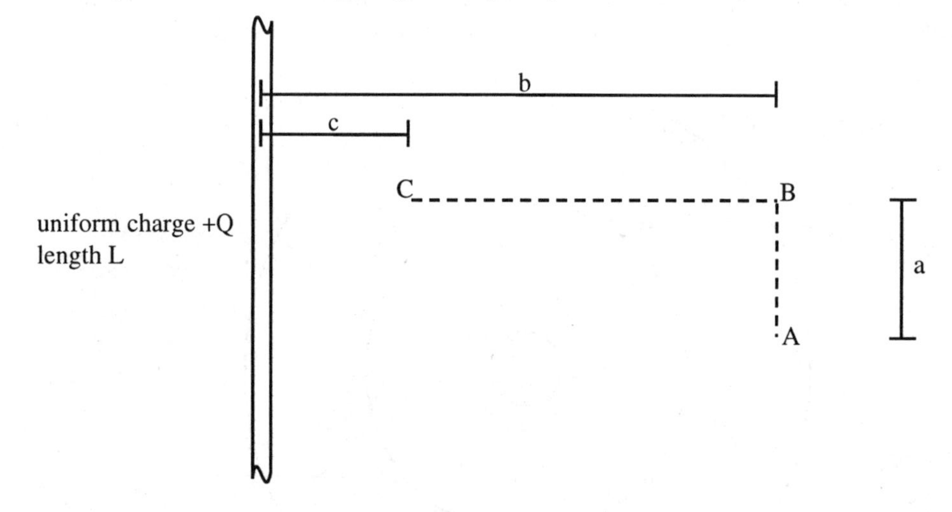

**Prob. 5 (55 pts):** Here is a bar magnet aligned east-west, with its center 16 cm from the center of a compass. The compass is observed to deflect 50 degrees away from north, and the horizontal component of the Earth's magnetic field is known to be $2\times10^{-5}$ tesla.

**(a, 5 pts)** Label the N and S poles of the bar magnet and explain your choice briefly.

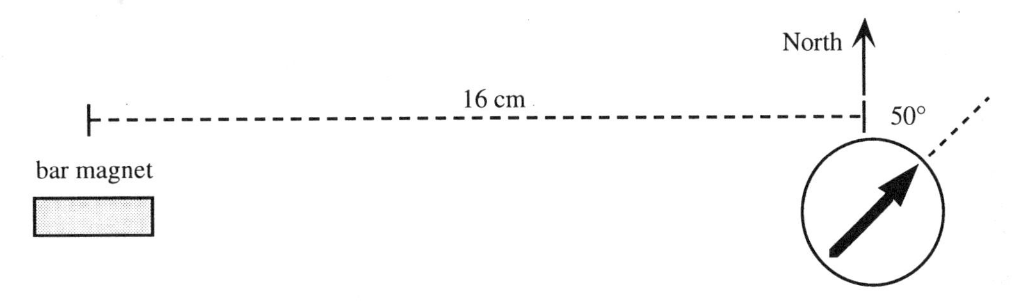


**(b, 10 pts)** Determine the magnetic moment of this bar magnet, including correct units. Use only formulas from the formula sheet. Show and explain all work.

You throw the bar magnet downward with its South end pointing down. Lying on the table is a nearly flat circular coil of 1000 turns of wire, with radius 5 cm. The coil is connected to an oscilloscope, which has a very large resistance.

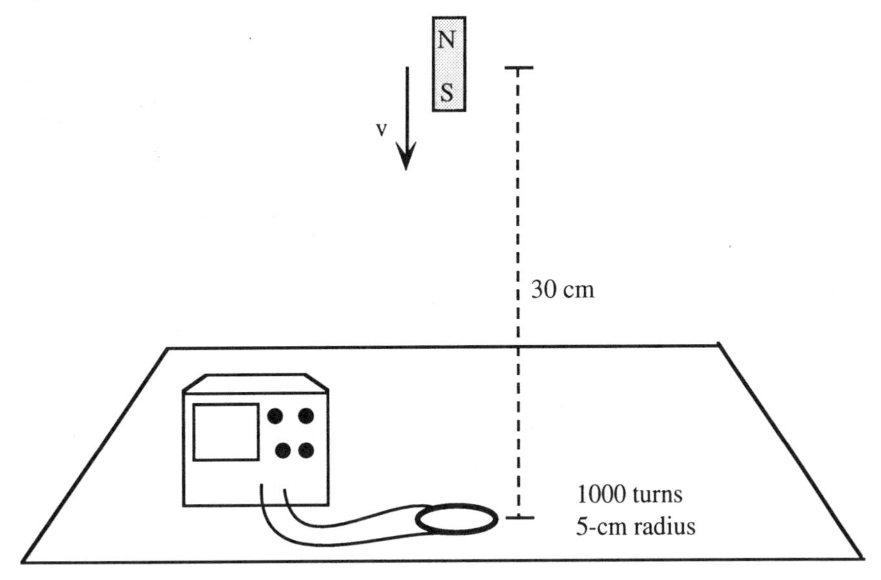


**(c, 10 pts)** On the diagram, show the pattern of non-Coulomb electric field in the coil. Explain briefly.

For part (d), use the value of μ that you calculated in part (b). If, however, you were *not* able to determine the magnetic moment in part (b), give a reasonable order-of-magnitude value based on your measurements of your own bar magnet, and use that value in part (d). Give correct units:

$\mu =$

**(d, 30 pts)** At the instant when the magnet has fallen to a height of 30 cm above the table, the oscilloscope indicates a voltage of magnitude 2 millivolts. What is the speed v of the magnet at that instant?

**Prob. 6 (40 pts):** Here is a circuit consisting of two 1.5-volt flashlight batteries and two nichrome wires of different lengths and different thicknesses as shown (corresponding roughly to your own thick and thin nichrome wires). The thicknesses of the wires have been exaggerated in order to give you room to draw the electric field inside the wires.

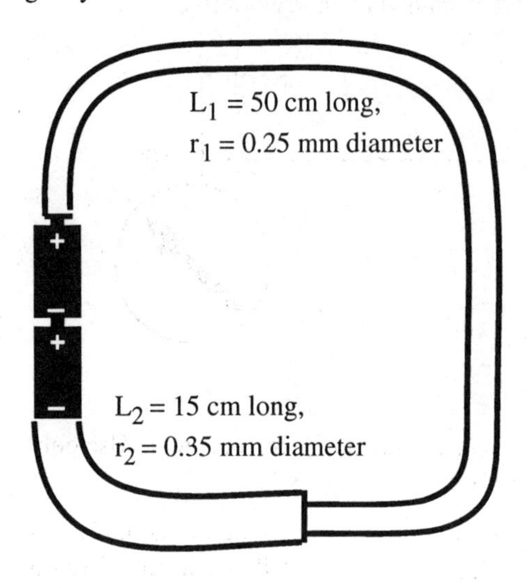


**(a, 10 pts)** Show the pattern of electric field throughout the wires, paying attention to appropriate relative magnitudes of the vectors that you draw. Explain briefly.

**(b, 20 pts)** Determine the magnitude of the steady-state electric field inside each nichrome wire.

**(c, 10 pts)** The electron mobility in room-temperature nichrome is about $7 \times 10^{-5} \dfrac{m/s}{N/C}$, and there are about $9 \times 10^{28}$ free electrons per cubic meter in nichrome. What is the conventional current through the thick wire, in amperes?

**Problem 7 (35 pts):** In the simple mass spectrometer shown below, positive ions are generated in the ion source. They are released, traveling at very low speed, into the region between two accelerating plates between which there is a potential difference $\Delta V_a$. In the region inside the dashed lines there is a uniform magnetic field B; outside this region there is no magnetic field. The gray line traces the path of one singly charged positive ion of mass M, which travels through the accelerating plates into the magnetic field region, and hits the ion detector as shown.

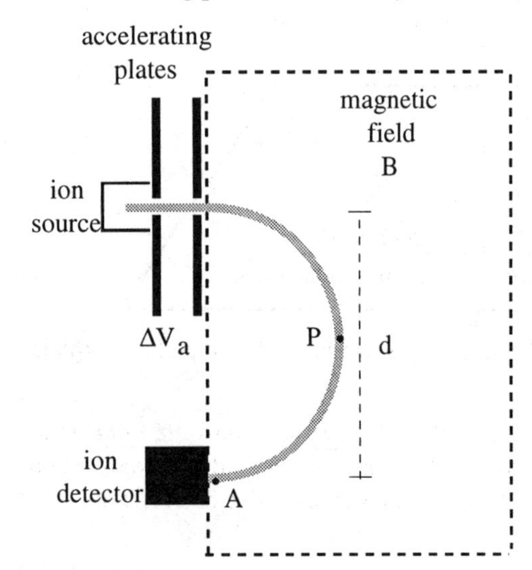


KNOWN QUANTITIES:

$\Delta V_a$: potential difference between accelerating plates

d: distance between entry slit and ion detector

M : mass of singly charged positive ion

**(a, 20 pts)** Determine the appropriate *magnitude* and *direction* of the magnetic field B, in terms of the known quantities listed above. *Use only formulas given on the formula sheet.* Explain all steps in your reasoning.

**(b, 15 pts)** At a time when an ion reaches location A, just before it enters the detector, what is the *magnitude* and *direction* of the magnetic field $B_{ion}$ at the location P, halfway along the trajectory between entry slit and ion detector, *due to this moving ion?* *Use only formulas given on the formula sheet.* Explain all steps in your reasoning.

**Formula sheet for Final Exam**

## CONSTANTS & MATH

$\frac{1}{4\pi\varepsilon_0} = 9\times10^9 \frac{\text{N-m}^2}{\text{C}^2}$

$\frac{\mu_0}{4\pi} = 10^{-7} \frac{\text{tesla-m-s}}{\text{C}}$

$c = 3\times10^8 \frac{\text{m}}{\text{s}}$

$\varepsilon_0 = 9\times10^{-12} \frac{\text{C}^2}{\text{N-m}^2}$

$e = 1.6\times10^{-19}$ C

$g = 9.8 \frac{\text{N}}{\text{kg}}$

$E = 3\times10^6 \frac{\text{N}}{\text{C}}$ breakdown of air

$m_{\text{electron}} = 9\times10^{-31}$ kg

$m_{\text{proton}} = 1.7\times10^{-27}$ kg

circumference of circle $= 2\pi r$

area of circle $= \pi r^2$

surface area of sphere $= 4\pi r^2$

chain rule: $\frac{d}{dx}\left[f(y)\right] = \frac{df}{dy}\frac{dy}{dx}$

## ELECTRICITY

$\int_{\text{closed surface}} \vec{E}\cdot\hat{n}dA = \frac{\sum q_{\text{inside}}}{\varepsilon_0}$

$\vec{E}_{\text{point}} = \frac{1}{4\pi\varepsilon_0}\frac{q}{r^2}\hat{r}$

$E_{\substack{\text{inside}\\ \text{insulator}}} \to \frac{E}{K}$

$\Delta V = V_f - V_i = -\int_i^f \vec{E}\cdot d\vec{l}$

$E = -\frac{\Delta V}{\Delta l}$

$\text{Work} = q\Delta V = \Delta\left(\frac{1}{2}mv^2\right)$

$V - V_\infty = \frac{1}{4\pi\varepsilon_0}\frac{q}{r}$

$\text{Power} = I\Delta V$

$\#/s = nAv$

$v = uE$

Ohmic: $I = \frac{\Delta V}{R}$

$\sum \Delta V_{\substack{\text{circuit}\\ \text{loop}}} = 0$

$\sum I_{\substack{\text{out of}\\ \text{circuit node}}} = 0$

## MAGNETISM

$\int_{\text{closed surface}} \vec{B}\cdot\hat{n}dA = 0$

$\int_{\text{closed path}} \vec{B}\cdot d\vec{l} = \mu_0\left[\sum I_{\text{inside}} + \varepsilon_0\frac{d\Phi_{\text{electric}}}{dt}\right]$

$\Phi_{\text{electric}} = \int_{\text{bounded surface}} \vec{E}\cdot\hat{n}dA$

$\vec{B} = \frac{\mu_0}{4\pi}\frac{q\vec{v}\times\hat{r}}{r^2}$

$d\vec{B} = \frac{\mu_0}{4\pi}\frac{Id\vec{l}\times\hat{r}}{r^2}$

$\mu = IA,\ B \approx \frac{\mu_0}{4\pi}\frac{2\mu}{x^3}$

$\text{emf} = \int_{\text{closed path}} \vec{E}\cdot d\vec{l} = -N\frac{d\Phi_{\text{magnetic}}}{dt}$

$\Phi_{\text{magnetic}} = \int_{\text{bounded surface}} \vec{B}\cdot\hat{n}dA$

$\text{P.E.} = -\vec{\mu}\cdot\vec{B}$

$\vec{F} = q\vec{E} + q\vec{v}\times\vec{B}$

$d\vec{F} = Id\vec{l}\times\vec{B}$

$\vec{E}_{\text{radiative}} = \frac{1}{4\pi\varepsilon_0}\frac{-q\vec{a}_\perp}{c^2 r}$, propagates in direction of $\vec{E}\times\vec{B}$

## MECHANICS

$\vec{F} = m\vec{a}$

$a_{\text{circular}} = \frac{v^2}{r}$

# CHAPTER 8

# FORMAL LABS

In addition to the many desktop experiments integrated with theory throughout the textbook, we have our students do two 2-hour formal labs that require more elaborate equipment. In the circuit lab the students use digital multimeters to study their own equipment in more detail, and to gain experience in the use of multimeters. In the magnetic induction lab the students not only gain hands-on experience with induction phenomena but also gain experience in the use of an oscilloscope.

The labs are structured in a form similar to the textbook, with in-line exercises on theory and experiment. The lab forms are collected at the end of the lab session—we do not require a formal lab report.


Circuit lab with multimeters 118

Magnetic induction lab 129

# CIRCUIT LAB WITH MULTIMETERS

## 1. INTRODUCTION

This lab experiment is intended to provide experience in using commercial multimeters for measuring and analyzing simple circuits. The accurate measurements possible with these devices complement your other experiments with circuits. The investigations include:

- energy conservation and current conservation
- ohmic and non-ohmic resistors
- internal resistance of a battery
- time dependence of a resistor-capacitor circuit

### *METERS*

You should have two multifunctional meters ("multimeters") that can serve as voltmeters for measuring potential differences, as ammeters for measuring current, or as ohmmeters for measuring resistances, depending on how the main switch on the meter is set.

The ammeter you will use in this experiment is functionally equivalent to the compass you used in your circuit experiments (although the actual construction of the ammeter is quite different). An ideal ammeter has very low resistance and should not affect the current you are trying to measure. Of course any real ammeter that is inserted into a circuit does have some resistance (unlike your compass, which was outside the circuit).

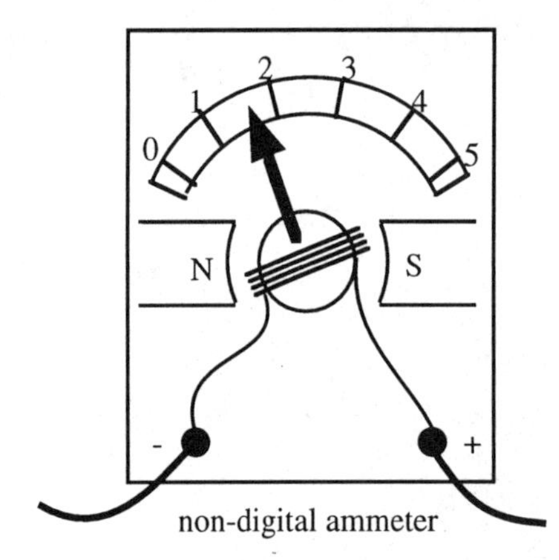

non-digital ammeter

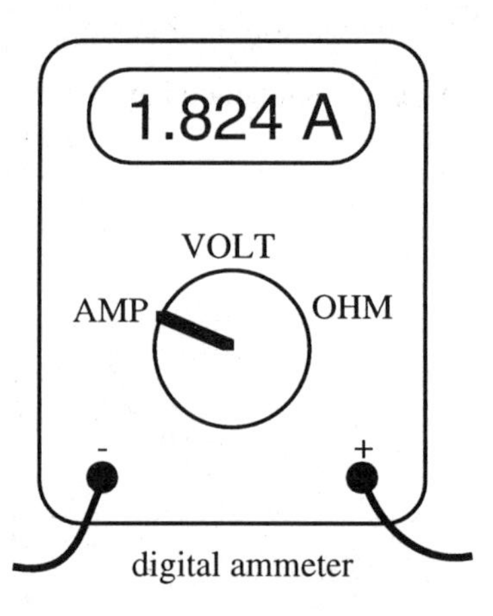

digital ammeter

The voltmeter used in this experiment can be thought of as a very sensitive ammeter in series with a very large (known) resistance R. When you connect the voltmeter between two points that have different potentials, a tiny current I runs through the voltmeter, and the amount of current is an indication of the potential difference ($\Delta V = RI$). In many practical situations this current is so small that attaching the voltmeter hardly affects the operation of the circuit.

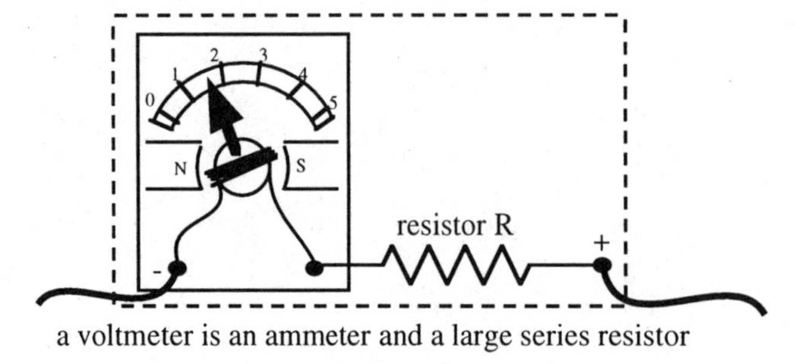

a voltmeter is an ammeter and a large series resistor

For operational details on the particular brand of meters you will use, see the accompanying sheet.

## 2. ENERGY CONSERVATION AND CURRENT CONSERVATION

In any circuit, energy conservation and current conservation yield loop and node equations:

$\Sigma(\Delta V) = 0$ around any loop

$\Sigma(I_{out}) = \Sigma(I_{in})$ at any node in steady state (charge conserved, not piling up anywhere)

You will verify by more accurate measurements than can be performed with your compass that these equations correctly describe a circuit. In the process you will learn how to use a commercial multimeter to measure currents and potential differences.

You are provided with a 47-ohm carbon resistor and a 100-ohm carbon resistor. The resistors are color-coded to indicate both their approximate resistance (the first three colored bands) and the tolerance or uncertainty in this approximate resistance (the fourth colored band):

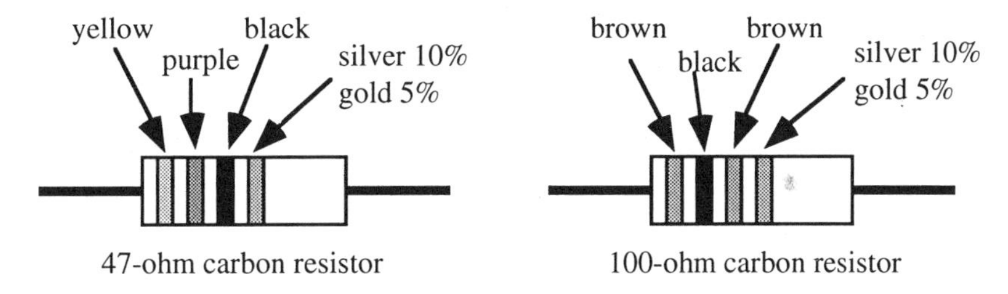


### Resistors in series

Consider a simple series circuit of 2 batteries, two resistors, and two ammeters:

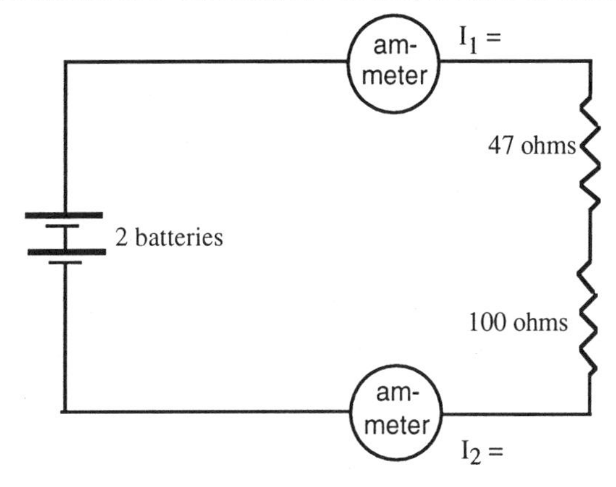


***Predict*** the currents $I_1$ and $I_2$ that would be measured by the two ammeters:

| $I_1$ = | $I_2$ = |
|---|---|

Next, assemble this circuit, but ***before connecting the two batteries***, have your instructor check your circuit. A wrong connection can blow a fuse or even damage an ammeter!

Connect the ammeters in such a way that the ammeters will read positive values. An ammeter will read a positive value if conventional current flows *into* the terminal marked "+". After your instructor has checked your circuit, connect the two batteries and **record the ammeter readings $I_1$ and $I_2$ on the diagram above**. Choose the ammeter scales to give maximum accuracy. Were your predictions about right?

| |
|---|

Reverse the connections to an ammeter to see what the ammeter reads when inserted "backwards."

Switch one of the ammeters to be a voltmeter (some multimeters may also require changing from ammeter sockets to voltmeter sockets). Describe and explain the new readings on the two meters:

Remove one of the ammeters from the circuit and switch it to be a voltmeter. For each part of the path around the circuit, measure and record on the diagram the potential difference. Include the sign of each potential difference. To measure a potential difference, simply touch the two leads of the voltmeter to the two points between which you want to measure $\Delta V$. A voltmeter will read a positive value if the "+" lead touches a higher potential than the "-" lead. Also record the current I.

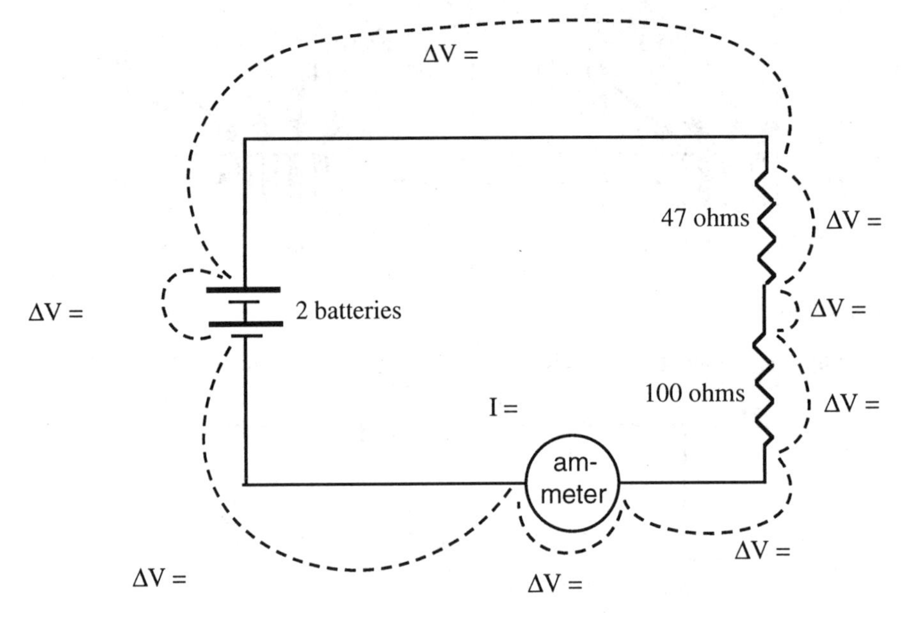


Energy conservation on a per-charge basis predicts that the individual potential differences should add up to zero for a round trip around the circuit. Do they?

Draw an approximate graph of potential (relative to the negative end of the battery) vs. position around the circuit, showing and labeling the various measured potential differences. Electric field is the (negative) gradient of the potential, so the slope of this graph should be steep where the electric field is large.

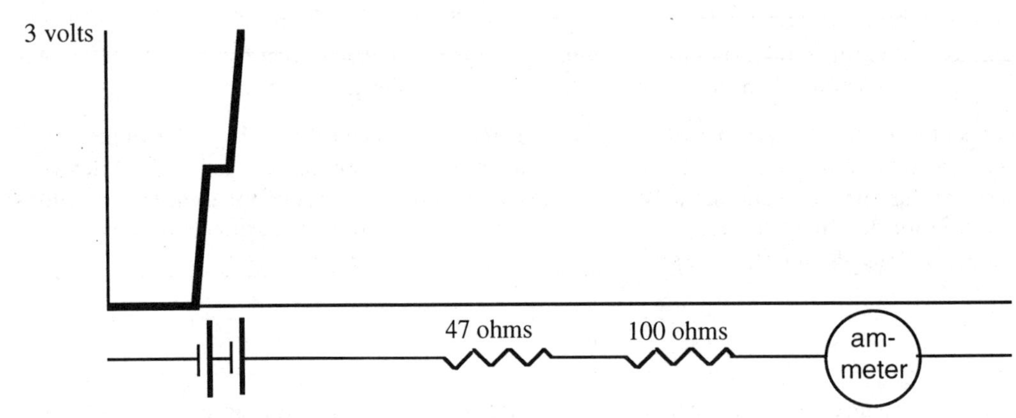

When you attach the voltmeter, you alter the circuit. Why doesn't the ammeter reading change?

Connect the voltmeter across the 100-ohm resistor again, and note the ammeter reading. Now switch the voltmeter to be a second ammeter. What happens to the current through the first ammeter? Why?

*From your measurements on page 3*, calculate the resistance of each of the following circuit elements:

$R_{\text{"47 ohm" resistor}} = \Delta V/I =$

$R_{\text{"100 ohm" resistor}} =$

$R_{wire} =$

$R_{ammeter} =$

A silver fourth band on a resistor indicates a manufacturing tolerance of plus or minus 10% (gold is 5%). This doesn't mean that the resistance fluctuates! It just means that when a batch of "100-ohm" resistors is manufactured, there is some variation around the approximate value of 100 ohms. A 10% uncertainty means that a "100-ohm" resistor is guaranteed to have a resistance between 90 ohms and 110 ohms. Because of this, trust your *own* measurement of the resistance, *not* the approximate value marked on the resistor.

**Using an ohmmeter**

*Take the circuit completely apart.* Choose an OHM setting on the multimeter, and connect the ohmmeter to a resistor *with nothing else connected to the resistor.* Use the ohmmeter to measure the resistances of the "47-ohm" resistor, of the "100-ohm" resistor, and of a long bulb (in a socket for convenience). How do the ohmmeter readings compare with your own previous measurements recorded above?

$R_{\text{"47-ohm" resistor by ohmmeter}} =$

$R_{\text{"100-ohm" resistor by ohmmeter}} =$

$R_{\text{long bulb}} =$

The reason you must remove a resistor from a circuit has to do with the way the ohmmeter works: a non-digital ohmmeter applies a small potential difference to a resistor and measures the current that flows, just as you did on page 3, so the ohmmeter is an active element. (A digital ohmmeter drives a small *current* through the resistor and measures the *potential difference*.) With the ohmmeter measuring a resistor, use your other meter to measure the potential difference across the ohmmeter:

$\Delta V_{ohmmeter} =$

### Resistors in parallel

Connect the 47-ohm carbon resistor and the 100-ohm carbon resistor in parallel with two batteries, with ammeters placed as shown. If you have only two ammeters, you will have to assemble two different arrangements in order to measure all three currents. Perhaps you can borrow a third ammeter from a neighboring group. **Be sure you do *not* connect an ammeter to the batteries without one or more resistors in series with the ammeter, because you might damage the ammeter!**

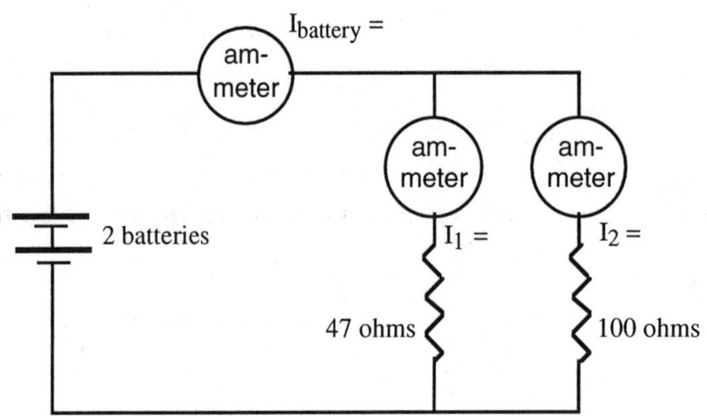


On the diagram, record the total current $I_{battery}$ and the currents $I_1$ and $I_2$ through the resistors. Is current conserved? (Is the current $I_{battery}$ equal to the sum of the currents $I_1$ and $I_2$ through the two resistors?)

With ammeters in place to read $I_{battery}$ and $I_2$ (through the 100-ohm resistor), note what happens to the ammeter readings when you disconnect the 47-ohm resistor. Considering what you observe in the ammeter readings, criticize the following statement: "A battery always puts out the same amount of current."

### 3. OHMIC AND NON-OHMIC RESISTORS

A resistor is said to be "ohmic" if the current I through the resistor is related to the potential difference $\Delta V$ across the resistor by the equation $I = \Delta V/R$, ***where the resistance R is a constant*** and does not vary with the voltage difference $\Delta V$. Other resistors whose resistance is *not* constant as a function of potential difference are called "non-ohmic resistors." We will investigate both kinds of resistors.

The approach is to place an ammeter *in series* with a resistor, and to place a voltmeter *across* the resistor. By using different numbers of batteries in series, you get pairs of values for I and $\Delta V$ which can be used to test whether the resistor is ohmic or not. Here is a schematic diagram of the circuit:

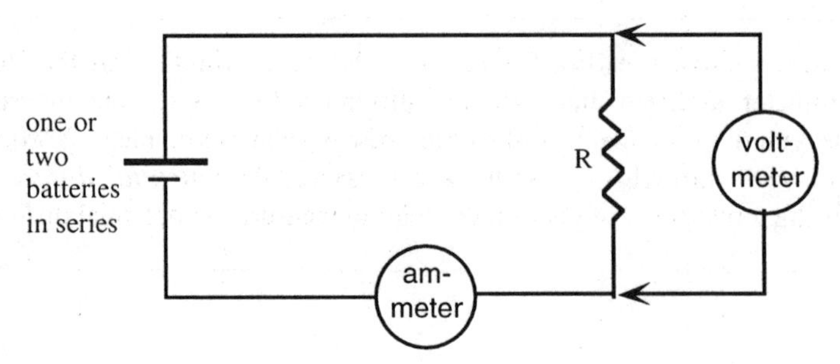

Place a 100-ohm resistor in the circuit shown above. Measure the potential difference $\Delta V$ across the 100-ohm carbon resistor and the resulting current I through the resistor with one battery and then with two batteries in series, and calculate the resistance. Also list your earlier ohmmeter measurements:

| $\Delta V$, volts | I, amperes or mA | R of "100-ohm" resistor |
|---|---|---|
| ohmmeter (bottom, p. 4): | I not known, but not needed; R read directly. | from p. 4, R = |
| 1 battery: | | $R = \Delta V/I =$ |
| 2 batteries: | | $R = \Delta V/I =$ |

Why do we say that a 100-ohm carbon resistor is an ohmic resistor?

Now replace the 100-ohm carbon resistor with a long bulb, and repeat the measurements:

| $\Delta V$, volts | I, amperes or mA | R of long bulb |
|---|---|---|
| ohmmeter (bottom, p. 4): | I not known, but not needed; R read directly. | from p. 4, R = |
| 1 battery: | | $R = \Delta V/I =$ |
| 2 batteries: | | $R = \Delta V/I =$ |

Evidently the long bulb is not an ohmic resistor. Briefly explain why its resistance depends on $\Delta V$, and explain the trend you observe:

**Nichrome wire**

Use an ohmmeter to measure the resistance of a long length of your *thin* nichrome wire, and the resistance of half that length. Theoretically, how should these two resistances compare with each other? Do your measurements agree? (Note: commercial ohmmeters often are not very accurate for resistances of just a few ohms.)

$R_{\text{full length of thin nichrome wire}} =$ $R_{\text{half length of thin nichrome wire}} =$

Use the ohmmeter to measure the resistance of the same lengths of your *thick* nichrome wire. Theoretically, how should these two resistances compare with each other? Do your measurements agree?

$R_{\text{full length of thick nichrome wire}} =$ $R_{\text{half length of thick nichrome wire}} =$

The thick wire has about twice the cross-sectional area of the thin wire. Theoretically, how should the resistance of the thick wire compare to the resistance of the thin wire? Do your measurements agree approximately? (The ratio of the areas may not be exactly 2 to 1.)

Connect a long length of the thin nichrome wire to two batteries in series, and use a voltmeter to measure $\Delta V$ between two points a constant distance $\Delta L = 5$ cm apart at various positions along the wire.

What are the numerical value and units of the quantity $\Delta V/\Delta L$?

What is the physical significance of the quantity $\Delta V/\Delta L$?

Why is $\Delta V/\Delta L$ the same at different places along the wire?

1 2 3 4 5 6 7 8 9 10 11 12 13 14 cm

## 4. INTERNAL RESISTANCE OF A BATTERY

This section requires an ammeter that can handle at least 10 amperes, which many commercial ammeters are not equipped to do. See the accompanying sheet for details about your equipment. If you *do not* have an ammeter capable of handling 10 amperes, skip this page.

If you *do* have an ammeter that can handle 10 amperes, assemble a circuit with one battery, the high-current ammeter, and a long thin nichrome wire in series. Place a voltmeter to read the potential difference across the battery.

***Do not make the final connections until your instructor has checked your circuit.***

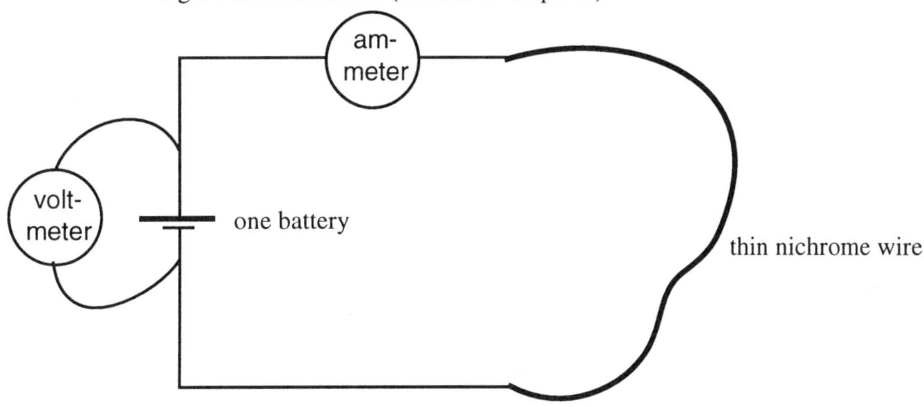


Vary the length of the thin nichrome wire by sliding the contacts. Be careful not to burn yourself on the hot wire!

Observe the behavior of the potential difference $\Delta V$ across the battery as the current I increases. Explain this qualitatively in terms of the fact that $\Delta V_{battery} = emf - r_{internal}I$:

If $I = 0$, $\Delta V_{battery} = emf - 0 = emf$. Measure the emf of your battery:

emf =

What is the maximum current you are able to observe? What does the voltmeter read? Use your data to calculate the internal resistance of the battery.

$I_{maximum}$ = $\Delta V_{battery}$ =

$r_{internal}$ =

The ammeter has some resistance, so $\Delta V_{battery}$ may not be zero. Remove the ammeter and nichrome wire. Connect several short copper wires in parallel across the battery. Now what does the voltmeter read, when there is a very large current running? Calculate the current that now runs through the battery.

$\Delta V_{battery}$ = $I_{battery}$ =

## 5. TIME-DEPENDENT VOLTAGE ACROSS A DISCHARGING CAPACITOR

You have seen that when a capacitor discharges through a light bulb the bulb starts out bright and then gets dimmer and dimmer. As the charge on the capacitor decreases, the voltage decreases ($\Delta V = Q/C$), and the current through the bulb decreases. It can be shown that when a capacitance C discharges through a resistance R, the voltage difference $\Delta V$ across the capacitor decreases exponentially with time according to

$$\Delta V = \Delta V_0 \, e^{-t/RC}$$

where $\Delta V_0$ is the initial voltage difference across the capacitor at time t = 0. Taking the natural logarithm of both sides of this equation yields

$$\ln \Delta V = \ln \Delta V_0 - (1/RC)t$$

$$\ln(\Delta V/\Delta V_0) = -(1/RC)t$$

Therefore, a plot of $\ln(\Delta V/\Delta V_0)$ vs. t should yield a straight line with a slope of -1/RC. RC is called the "time constant" of the circuit, the time at which the voltage difference has dropped to 1/e of its original value.

### Measurements

Fully charge your 0.47-farad capacitor by connecting it directly for a few seconds through clip leads to a single battery:

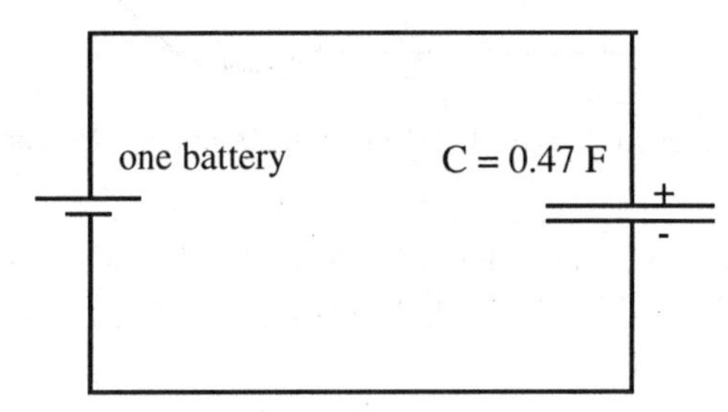


Disconnect the charged capacitor. Assemble the following circuit, but don't make the final connection to the capacitor until you are ready to start timing what happens. You will need a stop watch that measures in seconds. If you do not have a stop watch, the instructor will supply one.

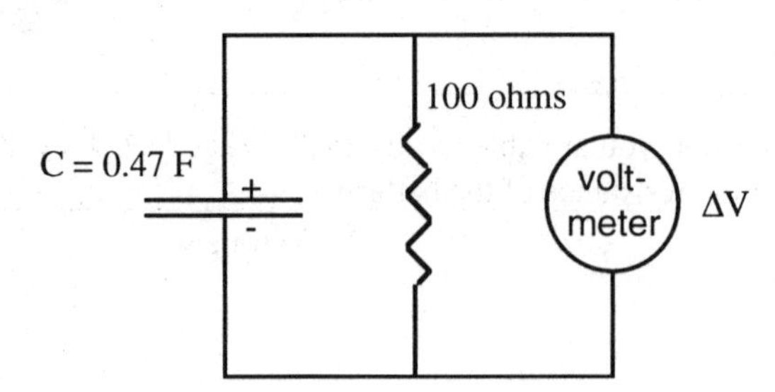

Record the potential difference across the capacitor every 10 seconds for 140 seconds, and after taking the data calculate $\ln\frac{\Delta V}{\Delta V_0}$ for each measurement:

| t, sec | ΔV, volts | $\ln\frac{\Delta V}{\Delta V_0}$ | t, sec | ΔV, volts | $\ln\frac{\Delta V}{\Delta V_0}$ | t, sec | ΔV, volts | $\ln\frac{\Delta V}{\Delta V_0}$ |
|---|---|---|---|---|---|---|---|---|
| 0 | | | 50 | | | 100 | | |
| 10 | | | 60 | | | 110 | | |
| 20 | | | 70 | | | 120 | | |
| 30 | | | 80 | | | 130 | | |
| 40 | | | 90 | | | 140 | | |

Plot $\ln\frac{\Delta V}{\Delta V_0}$ vs. t:

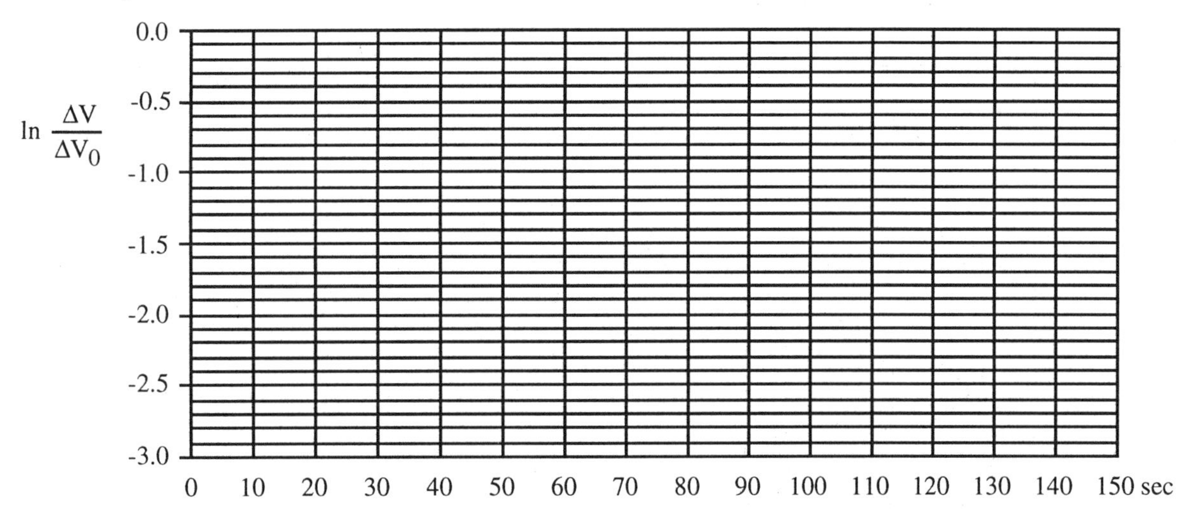


The slope of this line should be -1/RC. Use your own measured value for the resistance R of the "100-ohm" resistor, and determine C from the slope of the line. Is this within the 20% manufacturing tolerance of the stated value of 0.47 farad?

## Details on our particular multimeters

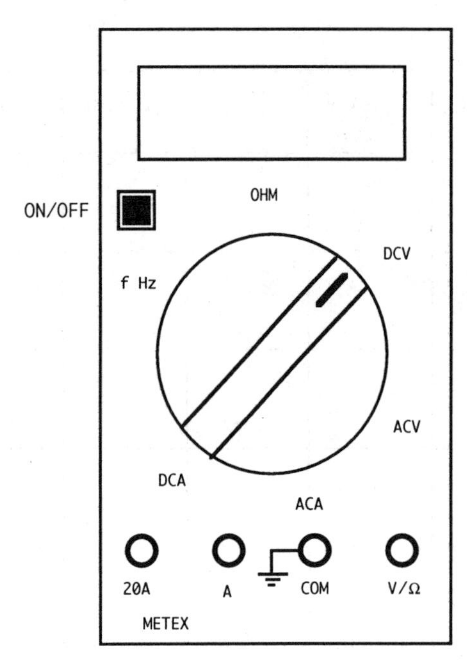


### Yellow Metex brand meter

Volts or ohms: connect to the COM and V/Ω sockets, and choose a DCV or OHM setting.

Currents: less than 0.2 amperes or 200 milliamperes (200 in the DCA section), connect to the COM and A sockets, and choose a DCA setting other than 20A.

**Currents greater than 0.2 amperes, up to 20 amperes: connect to the COM and 20A sockets, set to 20A in the DCA section. WARNING: Unless you connect to the 20A socket, a current of greater than 0.2 amperes will blow a fuse, which is hard to replace.**

Push in the ON/OFF button.

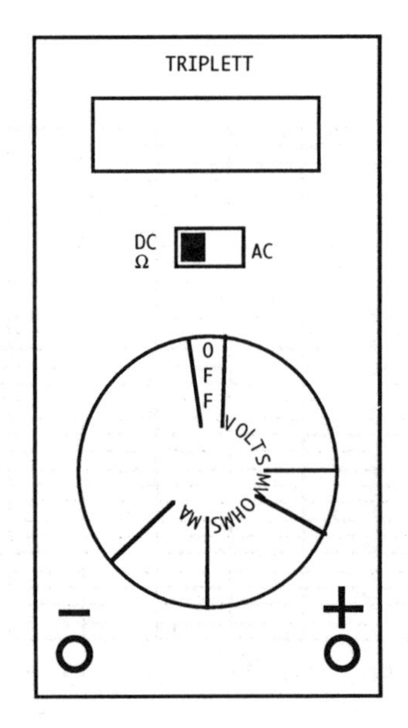


### Black Triplett brand meter

Set upper slider switch to DC/Ω ("direct current, ohms").

Rotate dial to choose "Volts" or "Ohms" or "MA" (milliamperes). You probably won't need to use "MV" (millivolts).

**WARNING: A current of greater than 2 amperes will blow a fuse, which is hard to replace.**

Choose the scales that have the lowest usable maximum value so that your measurements have the highest accuracy the meter can provide. A 200 volt setting lets you measure potential differences up to 200 volts, but if the potential difference of interest is only about one volt, you should use the 2 volt setting for greatest accuracy.

***WARNING: Unless you use the 20A option on the yellow METEX meter,***

***DO NOT attach an ammeter directly across the batteries!***

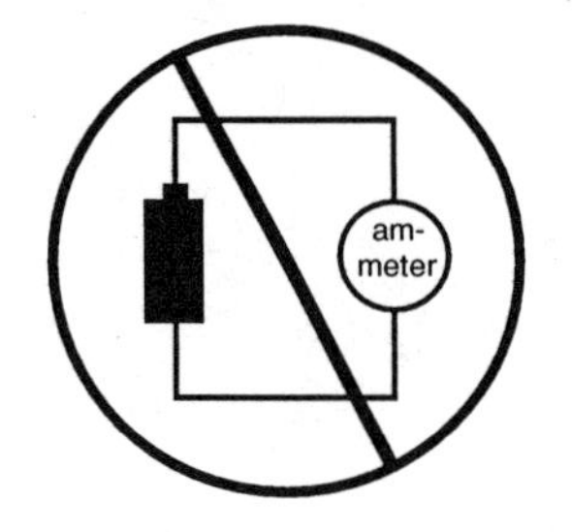


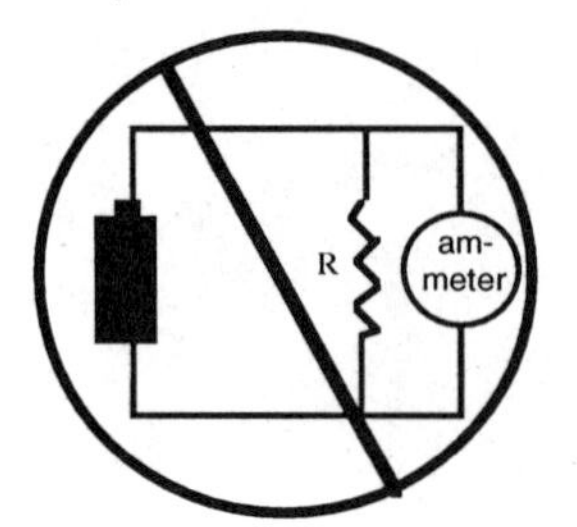

# MAGNETIC INDUCTION LAB

## Introduction

A changing magnetic flux through a surface generates an emf around the perimeter of that surface. We will use moving magnets and varying currents to study magnetic induction, and we will also get useful experience in using an oscilloscope, which is an important tool in engineering and scientific work.

## Setting up the oscilloscope and the AC power supply

Set the oscilloscope controls as follows:

Turn on the power.
Rotate the small red buttons on the center of the SWEEP and the VOLTS/DIV controls all the way clockwise until they click into "calibrate" mode. **Leave them calibrated at all times.**
Set the SWEEP TIME/DIV control to 1 mS (one millisecond per division).
Set the MODE switch to CH B (Channel B).
Set the TRIGGERING SOURCE to CH B.
Set the VOLTS/DIV for CH A and CH B to 1 V (one volt per division).
Set the AC-GND-DC switch for CH A and CH B to DC (*not* AC).

Set the controls on the AC (alternating-current) power supply (or signal generator or function generator):

Turn on the power.
Choose the 1KHz frequency selection.
Turn the frequency dial to 1 (to get 1KHz).
Set the SINE-SQUARE switch to SINE.
Set the amplitude control to the middle of its range.
Connect the output to the CH B input of the oscilloscope (see following note).

Note: The metal cases of the oscilloscope and of the power supply are connected to ground through the third pin of the power cord. Therefore, when connecting the output of the power supply to the oscilloscope, *"ground" on the power supply* ***must*** *be connected to "ground" on the oscilloscope.* The *black* connectors on the cables connect to the outer braid of the cable and to ground.

Adjust the amplitude knob *on the power supply* (not on the oscilloscope!) to produce a 2-volt amplitude, which is 4 volts peak-to-peak as shown on the diagram below. You may have to adjust the CH B POSITION control to move the display onto the screen. You may also have to adjust the TRIGGERING LEVEL to get an appropriate display. You should have a display on the screen that looks approximately like the diagram. If not, ask your instructor for help.

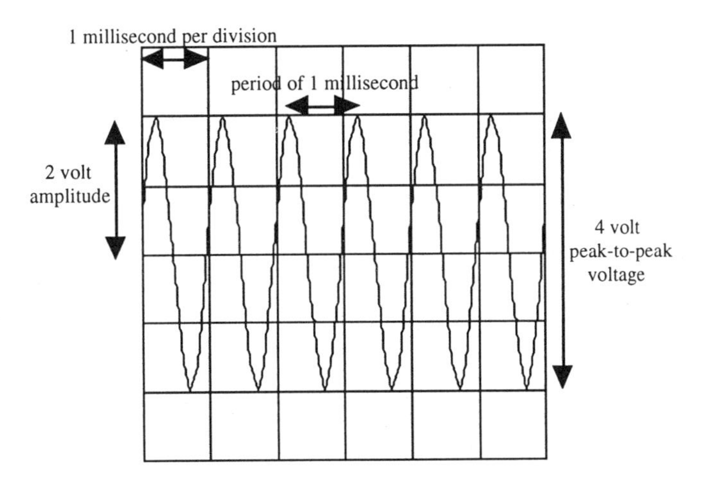

## Alternating current and voltage (AC)

An AC voltage $\Delta V(t) = V_0 \cos(\omega t+\alpha)$ is characterized by the *amplitude* $V_0$, *angular frequency* $\omega$ in radians per second, and *phase* $\alpha$. The *angular frequency* in radians per second is $\omega = (2\pi \text{ radians})/T = 2\pi f$, where T is the period in seconds and $f = 1/T$ is the frequency in hertz.

## The oscilloscope graphs voltage vs. time

The oscilloscope makes a graph of voltage on the y-axis vs. time on the x-axis, and it continually redraws the graph. The voltage is displayed by the vertical deflection of the electron beam, controlled by the VOLTS/DIV knob. When the VOLTS/DIV knob is set at 1 V, an input signal of 1 volt produces a vertical deflection of 1 division. In the figure on the previous page, the peak-to-peak voltage is 4 divisions or 4 volts, and the amplitude is 2 divisions or $V_0 = 2$ volts.

The *sweep circuit* of the oscilloscope generates a deflection voltage that moves an electron beam left-to-right across the tube face at an adjustable rate, controlled by the SWEEP TIME/DIV knob. When the SWEEP TIME/DIV knob is set at 1 mS (1 millisecond per division), it takes 0.001 sec for the trace to move horizontally from one vertical grid line to another. In the figure on the previous page, the period is T = 1 millisecond. What is the frequency in hertz?

f =

# Oscilloscope as DC voltmeter

Set the MODE switch to the DUAL setting, which displays simultaneously a voltage on Channel A (which isn't connected to anything yet) and the power supply voltage on Channel B. Adjust the vertical position of the CH A input until you see a horizontal line corresponding to zero voltage on Channel A.

Now connect the two wires leading to Channel A to a flashlight battery (with Channel B still connected to the power supply). What happens to the position of the horizontal line on the screen? What is the voltage across the battery?

Reverse the connection of Channel A to the battery. What happens to the position of the horizontal line on the screen?

The oscilloscope can be used as a sensitive DC voltmeter, and we'll use this capability to study the effects of a changing flux produced by a moving magnet.

## Changing magnetic flux due to a moving magnet

As shown in the diagram below, connect Channel A of the oscilloscope to one of the large coils that are provided. Leave the MODE switch set on the DUAL setting, which displays simultaneously both the coil voltage on Channel A and the power supply voltage on Channel B. Set the CH A VOLTS/DIV to 0.1 volts.

Move your bar magnet toward the coil, along the axis of the coil, with the N pole toward the coil. Observe the rise (positive coil voltage) or fall (negative coil voltage) of the horizontal line on the oscilloscope screen.

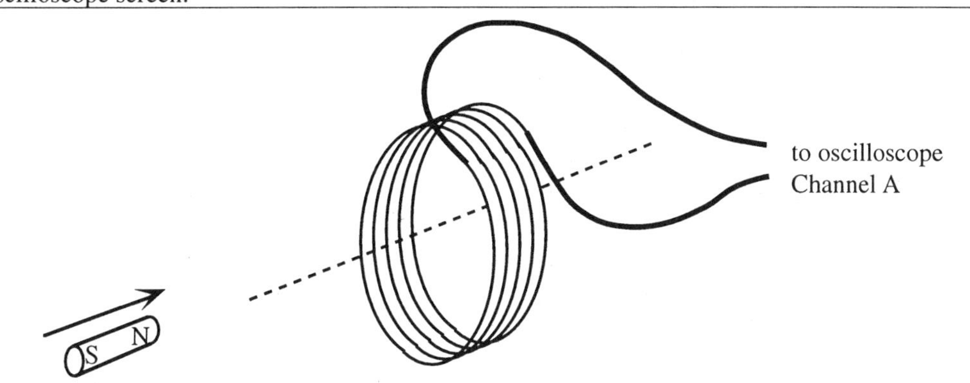


Using Lenz's law (or a right-hand rule for the non-Coulomb electric field), show on the diagram the direction of the induced conventional current in the wires.

As you move *toward* the coil, do you observe the coil voltage to be + or -?

As you move *away* from the coil, should the coil voltage be + or -? Is it?

Reverse the magnet, so the S pole is toward the coil.

As you move *toward* the coil, should the coil voltage be + or -? Is it?

As you move *away* from the coil, should the coil voltage be + or -? Is it?

Try moving the magnet slowly, instead of rapidly. How does this change the voltage? Why?

If you hold the magnet stationary, even if you are near the coil, why don't you see any coil voltage?

## Changing magnetic flux due to a varying current

Another way to produce a time-varying magnetic field and flux is to run alternating current through a coil, which we can do by connecting a coil to an AC power supply, which furnishes a sinusoidal voltage.

We will run a varying current through one coil to produce a varying magnetic field, and we will observe the effects of this varying magnetic field in a second coil. Leave Channel A of the oscilloscope connected to the coil. Turn off the AC power supply and disconnect it from the oscilloscope. Connect the *other* large coil in series with a 1000-ohm resistor to the power supply as shown, *making sure that one end of the resistor is connected to ground.*

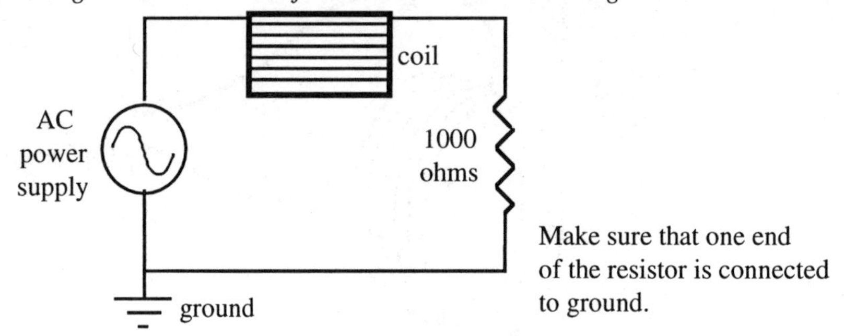


Connect Channel B of the oscilloscope across the resistor (to monitor the current through the resistor and coil). Note that one end of the resistor *must* be connected to ground. Since the metal chassis of the power supply and the metal chassis of the oscilloscope are both grounded through the wall sockets, connecting oscilloscope ground to the wrong point in the circuit would short-circuit some part of the circuit.

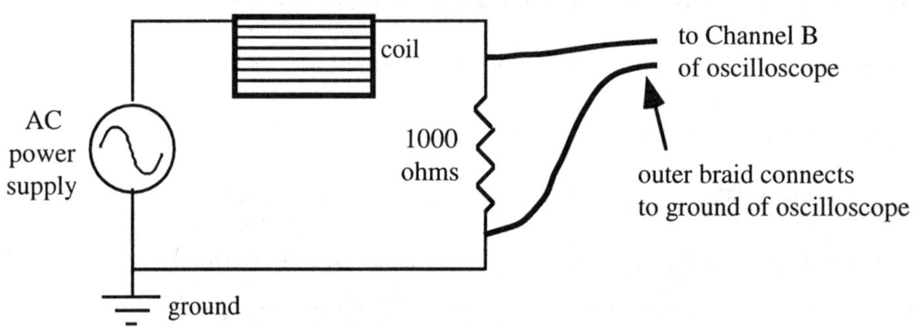


Place the two coils right next to each other, with their holes lined up with each other:

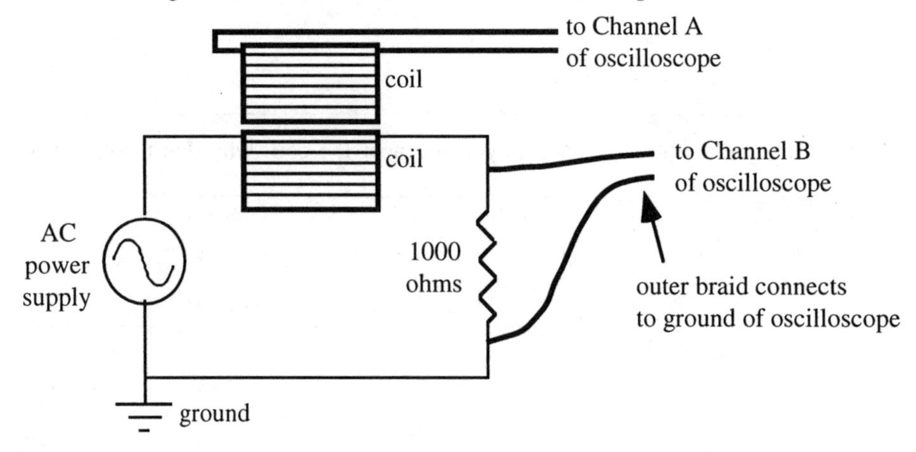


Turn on the AC power supply, which should still be set to supply a 1000-hertz sinusoidal voltage. Set the oscilloscope MODE switch in DUAL setting, which lets you display traces from Channel A and Channel B on the same time scale. Set the VOLTS/DIV to 0.5 for both CH A and CH B.

Adjust the output of the power supply to make the amplitude of the voltage across the resistor be 0.5 volt (one volt peak-to-peak), which means that the alternating current in the circuit has an amplitude of (0.5 volt)/(1000 ohms) = 0.5 milliampere (0.5 mA).

## Amplitude of induced emf

Adjust the VOLTS/DIV for CH A to obtain a signal coming from the coil that *isn't* connected to the power supply. Explain briefly why you observe some voltage across this coil, despite it not being connected to the power supply.

What happens to the amplitude as you move the two coils farther apart? Why?

What happens to the amplitude in this coil if you place one of the coils at an angle to the other one? Why?

## Phase of induced emf

Again place the two coils right next to each other, with their holes lined up with each other. Notice that the *current in the coil that is connected to the power supply* (which is proportional to the observed resistor voltage) and the *voltage across the other coil* are out of phase with each other. How many degrees (out of 360°) are the two sinusoids out of phase? (You can expand the time scale with the SWEEP control to get a more accurate measurement of this phase difference.)

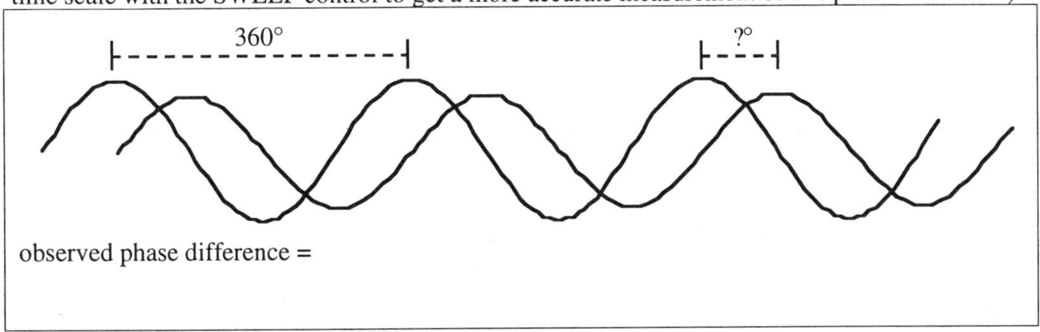


observed phase difference =

Explain the phase difference *quantitatively*. Hint: consider the current in the resistor (and the coil that is in series with the resistor) to be $I_0\cos(2\pi ft) = I_0\cos(\omega t)$, and think through exactly how this current in the first coil affects the second coil.

## Frequency dependence of magnetic induction

Next we will briefly study how the *frequency* of the AC power supply affects the induced voltage.

Again place the two coils right next to each other, with their holes lined up with each other. Make sure that the frequency is still set to 1000 hertz: adjust the frequency control *on the power supply* (not on the oscilloscope!) so that the period on the oscilloscope is exactly 1 msec. Set the VOLTS/DIV to 0.2 on both CH A and CH B.

Adjust the amplitude *on the power supply* (not on the oscilloscope!) so that the resistor voltage (CH B) is 0.4 volt ($I_0$ = 0.4 mA), or 0.8 volt peak-to-peak (4 divisions). Measure the amplitude of the coil voltage (CH A). It is often easiest to measure and record the peak-to-peak voltage, then divide by two.

| coil amplitude at 1000 Hz = |
|---|

Now change the power supply frequency to 500 hertz: adjust the frequency control *on the power supply* (not on the oscilloscope!) so that the period on the oscilloscope is exactly 2 msec.

Readjust the amplitude *on the power supply* (not on the oscilloscope!) so that the resistor voltage is again 0.4 volt ($I_0$ = 0.4 mA), or 0.8 volt peak-to-peak (4 divisions).

Again measure the amplitude of the coil voltage:

| coil amplitude at 500 Hz = |
|---|

Why does the amplitude of the coil voltage change with frequency, even though the amplitude of the current through the other coil is unchanged ($I_0$ = 0.4 mA)? Explain *quantitatively* the ratio of the coil amplitudes at 1000 Hz and 500 Hz:

We have seen that changing magnetic flux produces an emf in a coil. The flux change can be due to a moving magnet, a moving coil, or a changing magnetic field due to a changing current.

Alternating current (AC) is a convenient way to produce a time-varying magnetic field and flux.

## Self Inductance

Now we are going to study the phenomenon of "self-inductance," or the effects of the emf that a coil generates *in itself* when it is in a circuit. Remove the coil that is outside the resistor circuit, and disconnect the Channel A connections to that coil. Disassemble the resistor circuit. As shown below, reconnect the circuit so that Channel B is connected to the power supply, and Channel A is connected to the coil.

It is very important that all the ground connections be compatible as shown in the diagram. Note that one end of the coil *must* be connected to ground. Since the metal chassis of the power supply and the metal chassis of the oscilloscope are both grounded through the wall sockets, connecting oscilloscope ground to the wrong point in the circuit would short-circuit some part of the circuit.

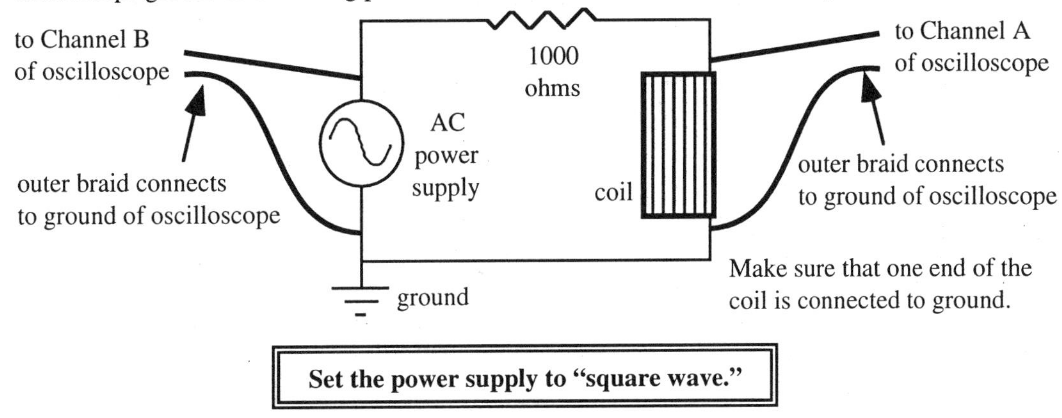


**Set the power supply to "square wave."**

Set the oscilloscope SWEEP TIME/DIV to 1 msec, and adjust the power supply frequency to 100 hertz so that the period on the oscilloscope is 10 msec (10 divisions). Set the VOLTS/DIV to 5 for both Channel A and Channel B. Set the MODE to CH B to observe the power supply voltage. Adjust the power supply amplitude to be 10 volts peak to peak, or two divisions.

Set the MODE to dual to observe both voltages. **Make an accurate sketch of what you see on the screen. Label this curve "coil voltage":**

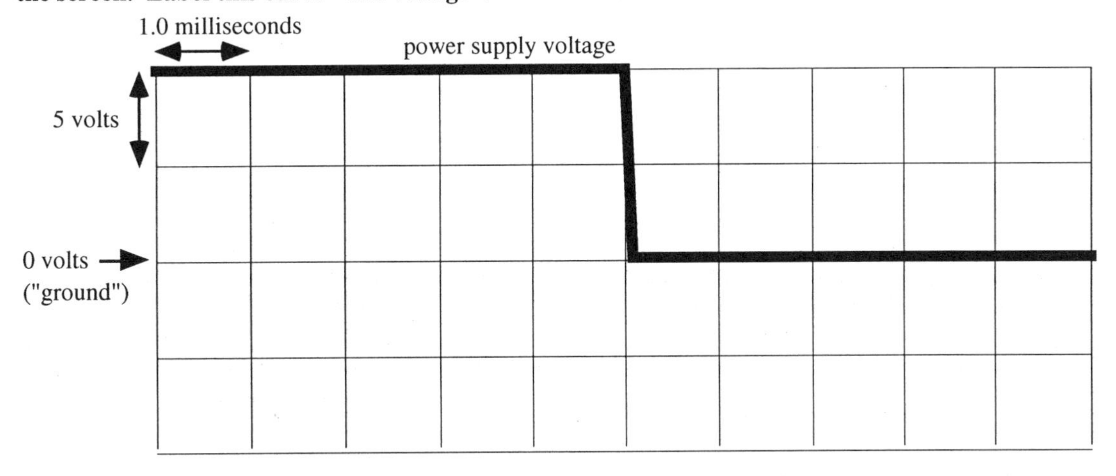


Assume that the energy-conservation loop equation is valid for this circuit *at every instant*: $\Delta V_{\text{power supply}} + \Delta V_{\text{resistor}} + \Delta V_{\text{coil}} = 0$. However, note that due to grounding constraints, while Channel B measures the potential rise across the power supply going clockwise around the circuit, Channel A measures not the potential drop across the coil going clockwise, but the potential rise going counter-clockwise. When Channel B measures +10 volts ($\Delta V_{\text{power supply}}$) and Channel A measures +9 volts ($-\Delta V_{\text{coil}}$), the potential drop across the resistor is 1 volt. Plot some dots for $\Delta V_{\text{resistor}}$ according to the energy-conservation loop equation, and **carefully connect the dots to graph the voltage across the resistor as a function of time. Label this curve "resistor voltage".**

There is no steady state in this circuit: there are continual rearrangements of the surface charges on the wires and circuit elements, which implies slight deviations from current conservation in a series circuit. However, it only takes a few nanoseconds to completely rearrange the surface charges, whereas the period of a 100-hertz square wave is *ten million* nanoseconds (10 milliseconds), so there is plenty of time for the surface charge to readjust. Despite the non-steady-state character of this circuit, current conservation as well as energy conservation can be considered to be valid at every instant.

A changing current dI/dt in the coil makes a changing flux NdΦ/dt *inside the coil itself* which is proportional to dI/dt. This produces a voltage *across the coil itself* proportional to dI/dt: $\Delta V_{coil}$ = -LdI/dt, where L is a constant and is called the "inductance" of the coil. The minus sign reflects the fact that the coil voltage is numerically equal to the emf developed in the coil, which opposes the change in the current.

On the graph on the previous page, ***qualitatively*** **sketch and label dI/dt**. (Note that I = $\Delta V_{resistor}$/R, and don't worry about the vertical scale of your sketch.) Is your graph of dI/dt proportional to $\Delta V_{coil}$?

The coil voltage in an RL circuit drops to 1/e (= 0.37) of its maximum value in an amount of time equal to L/R (see Chapter 13 in your textbook). Try changing the resistance, and see how this affects how rapidly the voltage across the coil drops to 1/e of its maximum value. Try a range from the smallest available resistance up to a maximum of 10KΩ. If you increase the resistance R, does the coil voltage drop by a factor 1/e faster or slower?

increase R; coil voltage drops by 1/e faster or slower?

On your graph of R = 1000 ohms, find the time when the coil voltage has dropped to 1/e (= 0.37) of its maximum value, and use this time (= L/R) to determine the inductance L of the coil in "henries" (volt-sec/ampere):

L =

We have seen that a coil (an "inductor") acts in a circuit in such a way that there is an emf and a corresponding potential difference across the inductor that is proportional to the rate of change of the current: $\Delta V_{coil}$ = -LdI/dt, where L is the "inductance" in henries. This "self-inductance" effect is due to the varying magnetic flux produced by the varying current in the coil.

## Oscillations in LC circuits

When a charged capacitor is connected to an inductor, the current in the circuit oscillates. This oscillation will go on forever unless there is some resistance in the circuit, in which case the oscillations will die out after a time. Oscillating circuits are common and extremely important. Your AC power supply is an example of such a circuit.

Replace the 1000-ohm resistor with a 0.001-µf (microfarad) capacitor. To simulate the effect of repeatedly flipping a switch to connect a charged capacitor to an inductor, leave the power supply option on "square wave," and set the power-supply frequency to 50 hertz. This has the effect of placing a charge on the capacitor 50 times a second, which starts oscillations in the circuit.

Temporarily change the SWEEP TIME/DIV to a slow enough rate that you can see the slow square wave with oscillations superimposed. Then set the SWEEP TIME/DIV to 0.1 msec or shorter time to observe the oscillations in detail. The oscillations die away because there is some resistance in the circuit. In particular, the large coil has significant resistance. If there were less resistance, the oscillations would die away less rapidly.

What is the period and frequency of the oscillation?

| period T = |
|---|
| frequency f = |

It can be shown that this "natural" frequency of an "LC" circuit is $f = \frac{1}{2\pi\sqrt{LC}}$ (see Chapter 13 in your textbook). Given your measurement of the frequency f and the known capacitance C = 0.001 microfarad, what is the inductance L of the coil? How does this compare with your earlier value for L? (Note that the resistance of the "1000-ohm" resistor isn't exactly 1000 ohms, and the rating of the "0.001-µf" capacitor is even less accurate. Moreover, the turns of the coil have some capacitance with respect to each other. So this is not a precision comparison.)

| L = |
|---|

## Resonance in LC circuits

Set the power-supply frequency to the natural frequency f of the LC oscillations you just observed (you may have to change the frequency range switch in order to reach the desired frequency). You may also have to adjust the Channel A VOLTS/DIV to get the rather large coil voltage on the screen.

> Be sure to switch the power supply from square wave back to sinusoidal voltage.

Vary the power-supply frequency lower and higher than the natural frequency. Qualitatively, what do you observe about the *amplitude and phase difference* of the coil voltage as you vary the frequency lower and higher than the natural frequency?

frequency dependence of amplitude:

frequency dependence of phase difference:

This strong dependence on frequency is called "resonance." (If you don't see a strong dependence on frequency, check with your instructor: you must not be near the natural frequency.)

It is an extremely important property of sinusoidally oscillating systems that they give a large response to applied frequencies near their natural oscillation frequency, and a very small response to applied frequencies far from their natural frequency. The smaller the resistance in the circuit, the sharper the resonance; that is, you get a bigger response and the response falls off more rapidly as you move away from the natural oscillation frequency.

Set the power-supply frequency to give the biggest coil voltage you can get. Record the amplitude of the power-supply voltage and the amplitude of the coil voltage:

power-supply amplitude:

coil amplitude:

Isn't it surprising that the coil amplitude is *huge* compared to the power-supply amplitude? (If it isn't, check with your instructor: you must not be near the natural frequency.)

To understand how this can be, imagine pushing somebody in a swing. If you time your pushes to match the natural frequency of the swing, you can build up and then maintain a huge amplitude of the swing with very small-amplitude pushes. In your resonant circuit, the oscillation needs only little pushes *at the right times* from the power supply to make up for resistive losses, and the amplitude of the oscillation can be much larger than the amplitude of the power supply. To satisfy the energy-conservation loop equation the voltage across the capacitor has to be nearly the opposite of the voltage across the inductor, so that the sum of these two large voltages adds up to the small power-supply voltage.

## Quantitative flux change

If you have extra time, the following optional experiment and analysis will help you better understand the quantitative relationship between emf and the rate of change of magnetic flux.

As on page 3, connect a coil to Channel A of the oscilloscope, and connect Channel B to the power supply in order to trigger the oscilloscope trace. Set the CH A VOLTS/DIV to 0.1 volts.

Estimate the flux change due to moving your bar magnet, based on the fact that $B_{magnet} = (\mu_o/4\pi)(2\mu/x^3)$ on the axis (far from the magnet). For the magnetic moment $\mu$ of your bar magnet use the value you determined in Chapter 11 of your textbook. *Roughly* estimate the amount of flux passing through *one turn* of the coil when the magnet is 10 cm from the coil, along the axis:

$\Phi_{10\text{ cm away}} = \sum \vec{B} \bullet \hat{n}\Delta A \approx$

State the simplifying assumptions you had to make:

Estimate the time it takes for you to quickly move the magnet from a position 10 cm from the center of the coil to a position about 30 cm away. Use a stopwatch and count how many repetitive moves you can make in a few seconds.

$\Delta t =$

The magnetic flux when you are 30 cm away is negligible compared to the flux when you are only 10 cm away, due to the rapid $1/x^3$ decrease in magnetic field of your magnet. So when you move the magnet between these two positions, the flux change $\Delta\Phi_{one\text{ turn}}$ is approximately equal to the magnitude of $\Phi_{10\text{ cm away}}$.

What voltage do you expect to see on the oscilloscope when you move the magnet between positions 10 cm and 30 cm distant from the coil? (You need to know N, the number of turns in the coil.) What do you actually observe?

$|\Delta V_{expected}| = N|\Delta\Phi_{one\text{ turn}}/\Delta t| =$

$|\Delta V_{observed}| =$

# CHAPTER 9

# SOLUTIONS TO SELECTED HOMEWORK PROBLEMS

Here we provide solutions and a few sample grading keys for selected homework problems. In our own course, solutions to assigned problems were posted only after students had turned in their work.


| | |
|---|---|
| Chapter 1 solutions | 142 |
| Chapter 2 solutions | 144 |
| Chapter 3 solutions | 147 |
| Chapter 4 solutions | 151 |
| Chapter 5 solutions | 159 |
| Chapter 6 solutions | 160 |
| Chapter 7 solutions | 165 |
| Chapter 8 solutions | 168 |
| Chapter 9 solutions | 178 |
| Chapter 10 solutions | 182 |
| Chapter 11 solutions | 186 |
| Chapter 12 solutions | 191 |
| Chapter 13 solutions | 198 |

## Chapter 1 solutions

**RQ1-5:**

Total charge of proton is +1e, which can be made with (+2/3 +2/3 -1/3)e, so uud (two u quarks and one d quark). Total charge of neutron is 0, which can be made with (+2/3 -1/3 -1/3)e, so udd (one u quark and two d quarks). There don t seem to be any other combinations of three quarks that can be used to make a proton and neutron.

**RQ1-7:**

Two kinds of evidence for electric interactions passing through other objects: Seems like excess charge on a U tape is on the sticky side, but it interacts when you approach the slick side (page 3). Also we tried observing repulsion of two U tapes with a piece of paper between (page 23), but it was hard to see because the U tapes are attracted to the paper, which opposes the repulsion.

**RQ1-8:**

Results consistent with Newton s third law: forces equal and opposite.

**HW1-1: (Grading key: 20 pts)** Amount of charge on a tape

1-1a (4 pts): describe approach

1-1b (4 pts): diagram -- should be clear and clearly labeled

1-1c (10 pts): physically correct analysis, Q in electron charges and coulombs

Typical values for tape 20 cm by 1.2 cm: $Q = 10^{-8}$ C, $6\times10^{10}$ e

-2 if do not explicitly state simplifying assumption of point charges

-1 if do not explicitly state simplifying assumption that $Q_1 = Q_2$

1-1d (2 pts): check result is reasonable

**HW1-2: (Grading key: 5 pts )** Fraction of molecules affected

$$\frac{(20\text{cm})(1.2\text{cm})}{\left(3\times10^{-8}\text{cm}\right)^2} = 2.7\times10^{16} \text{ molecules on surface}$$

$$\frac{6\times10^{10}\text{electrons}}{2.7\times10^{16}\text{molecules}} \approx 2\times10^{-6}, \text{ or about one in 500,000.}$$

-2 for simple arithmetic error

**HW1-4: (Grading key: 20 pts)** Electric Field Hockey

5 pts (a) Screen print of level 5 solution

5 pts (b) Screen print of level 5 with extra charge that has a big effect

2 pts (c) Points where acceleration is large

-1 if errors

-1 if don't explain (correctly) how you can tell from looking at trail
(dot spacing changes, trajectory curves)
("large force" is not ok)

2 pts (d) Points where acceleration is small

-1 if errors

-1 if don't explain (correctly) how you can tell from looking at trail
(dot spacing constant, straight line motion)
("small force" is not ok)

2 pts (e) Point where force & velocity differ in direction
(all or nothing)

2 pts (f) How do answers to c & d illustrate rapid falloff of Coulomb force
(should say large acceleration only very near a charge)

2 pts (g) Why did distant charge have an effect
(very small force, but acting over sufficient time/distance to produce an effect)

-2 if think charge had an effect only when particle got near it

**HW1-6:**

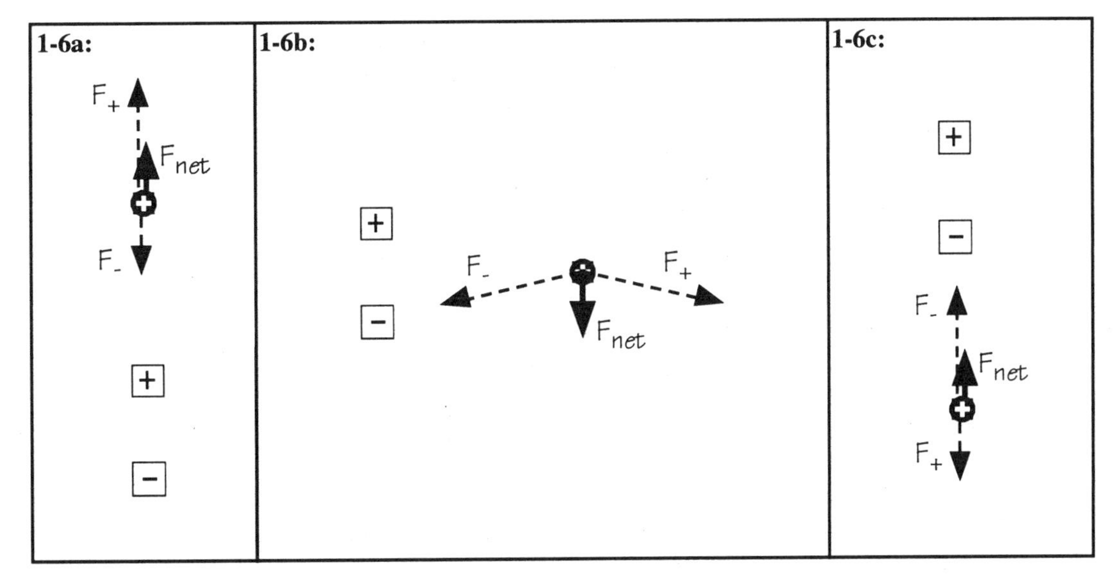

## Chapter 2 solutions

**RQ2-3:**

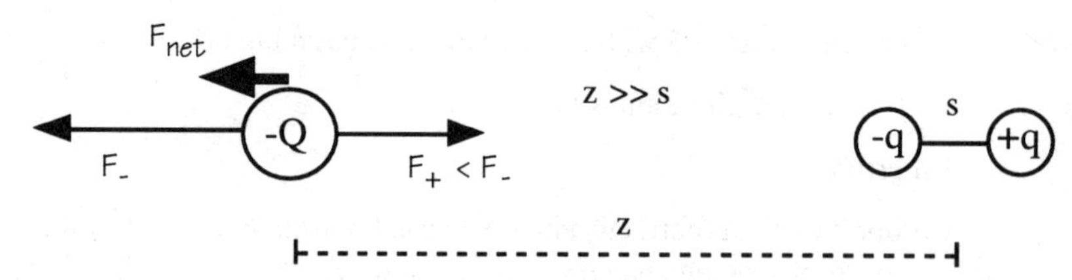


**RQ2-4:** Force by permanent dipole on point charge goes like $1/d^3$ (see section 2.1), so doubling the distance will reduce the force by a factor of $1/2^3 = 1/8$.

**RQ2-7:** The positive charge also *repels* the positive side of the molecules, though with a smaller force, so that the *net* force is an attraction.

**HW2-2: (with grading key: 20 pts)**

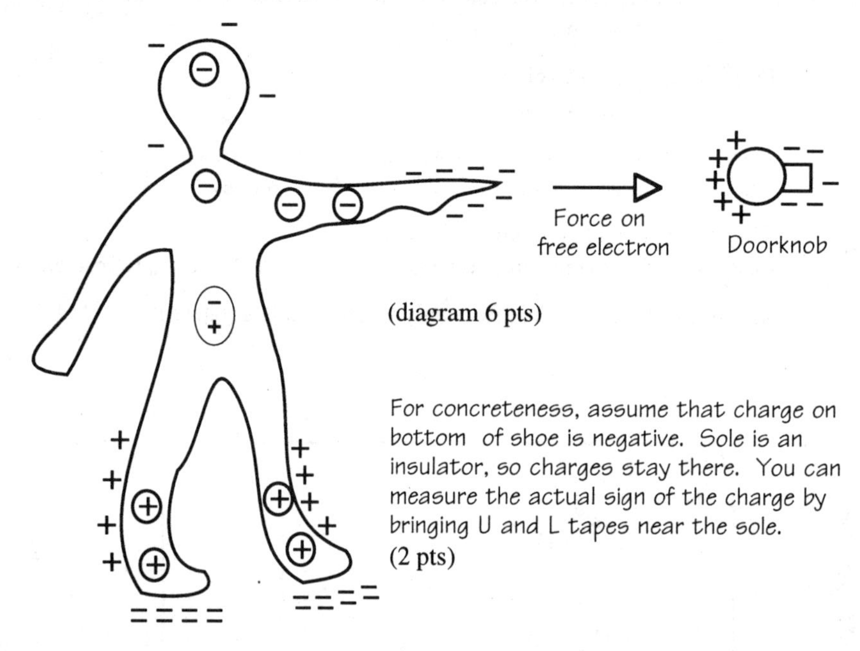


(diagram 6 pts)

For concreteness, assume that charge on bottom of shoe is negative. Sole is an insulator, so charges stay there. You can measure the actual sign of the charge by bringing U and L tapes near the sole. (2 pts)

The body polarizes, not only individual molecules, but flow of $Na^+$ and $Cl^-$ ions on the skin (salt water layer) and in the blood. So finger is negatively charged. (4 pts)

So doorknob polarizes as shown, and the charge separation on the doorknob, as well as the negative charge on the finger, exert a force to the right on a free electron in the intervening air. (4 pts)

As long as you are not very close to the doorknob, this force is not large enough to accelerate a (rare) free electron in the air sufficiently to give it enough energy to ionize an air molecule when it collides with one. So no spark yet. As you get closer, the doorknob and your finger polarize each other more. The force on a free electron in the air gets larger, until finally the electron accelerates enough between collisions with air molecules to have enough energy to rip an electron off an air molecule. This starts a chain reaction, ionizing the air a lot. (2 pts)

Electrons and ions now move through the air toward the doorknob and your finger. Sometimes an ion recombines with an electron in the air and emits light, hence the visible spark. Electrons flow onto the doorknob, charging it more and more negative. Positive ions ($N_2^+$ and $O_2^+$) hit your finger, and there are chemical reactions in the salt water layer. Your finger gets less negative. Eventually a free electron is not accelerated enough to ionize an air molecule, and the spark stops. (2 pts)

**HW2-7:**

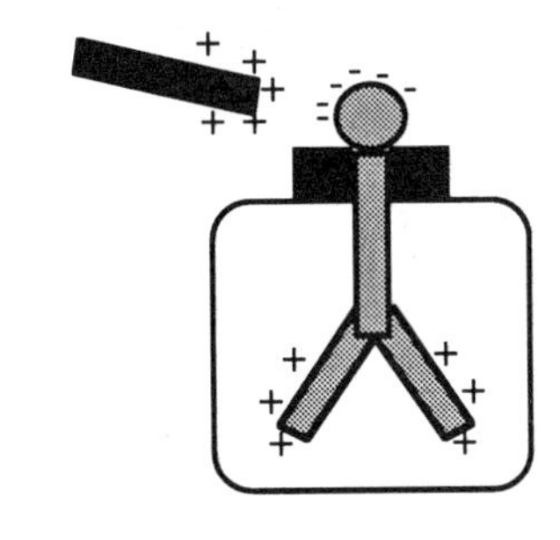

Positive glass rod attracts electrons toward top of metal, leaving deficiency below on both gold foils, which repel each other.

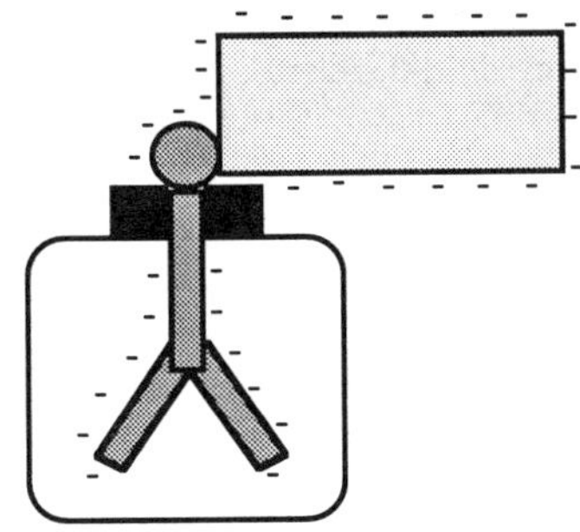

Assume that the metal block was charged negative. In contact, charge spreads over everything, including both foils, which repel each other.

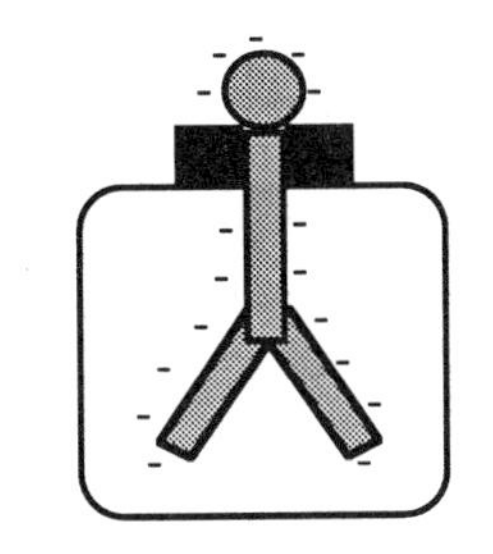

Foils stay spread apart when removed from metal block; still charged.

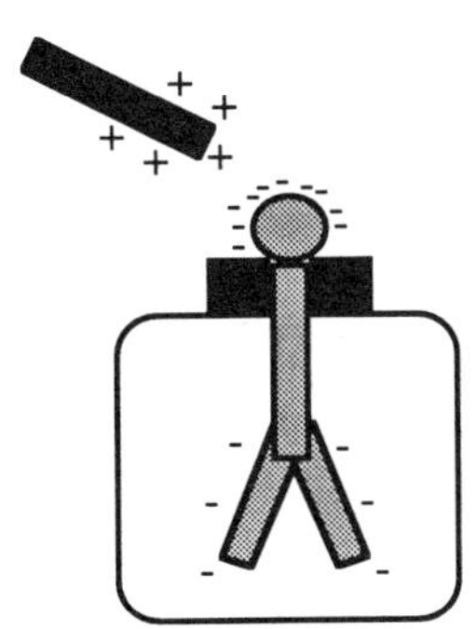

Positive glass rod attracts some of the excess electrons to the top, reducing the charge on the gold foils, so the foils don't repel each other as much.

If we had assumed the metal block was positive, approaching the positive glass rod would have made the gold foils more positive, and the foils would have moved farther apart.

**HW2-9:** We will present particular solutions, but these are *not* the only valid schemes.

(a)

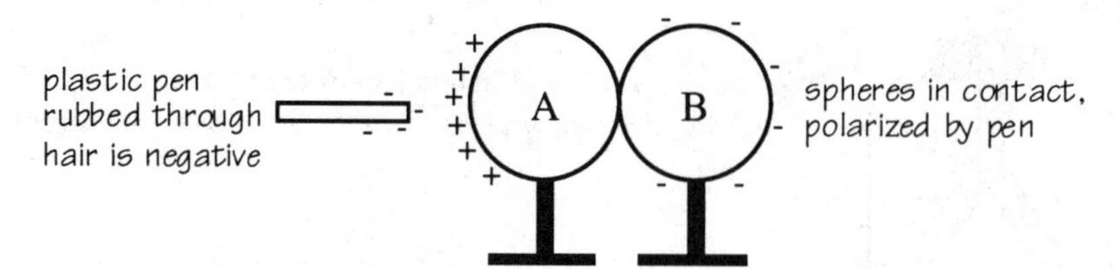


While keeping the pen in place, separate the two spheres.
They are charged equally and oppositely, since for each electron that moved from A to B, there was left behind one electron deficiency.

(b)

plastic pen
rubbed through
hair is negative

A B

touch here,
neutralizing
this region

A B

remove finger,
so net charge
of two spheres
is positive

remove pen

A B

positive charge
distributed equally,
since the two spheres
are identical

Now simply separate the two spheres: they are charged equally positive.

Another way to do this is to charge one sphere by induction, move hand and pen away, and bring the two spheres in contact with each other to share the positive charge equally.

It is not really adequate to charge one sphere by induction, then charge the other by induction. The amount of charge deposited on a sphere when charged by induction depends on exactly how close the pen is to the sphere. However, after charging both spheres by induction, you could touch the two spheres together and let the positive charge be distributed equally between the two spheres.

## Chapter 3 solutions

**RQ3-3:** Here are a couple of ways to arrange two point charges in order to make the electric field be zero at some location:

**RQ3-7:** Consider moving a point charge to twice the distance from a permanent dipole, in which case the force that the dipole exerts on the point charge drops by a factor of $1/2^3 = 1/8$:

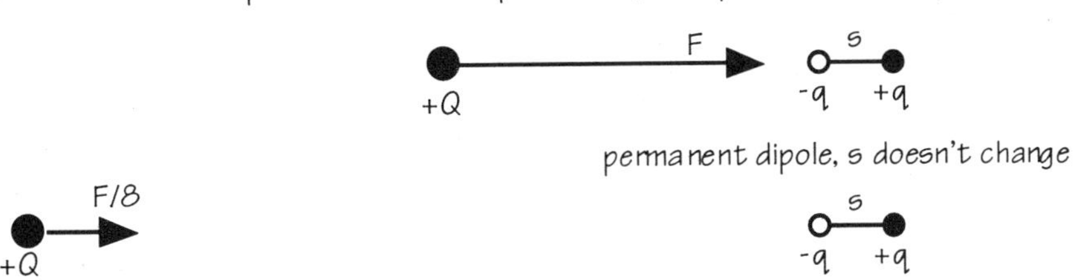


Now consider moving a point charge to twice the distance from an induced dipole, in which case the dipole separation drops by a factor of 1/4 (field of the point charge drops by 1/4), so the force on the point charge drops by a factor of (1/8)(1/4) = 1/32:

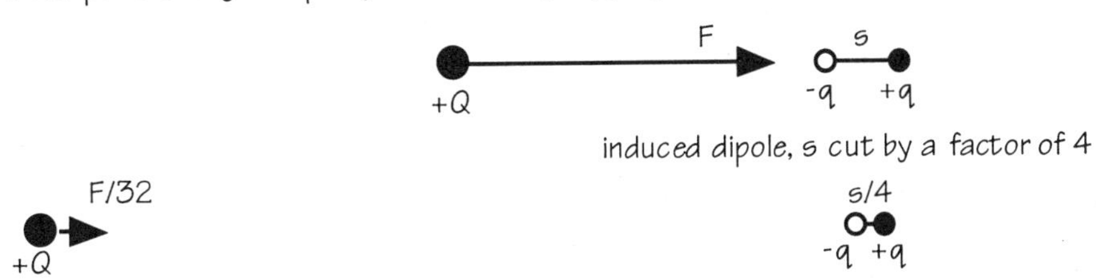


So the key point is that a dipole makes a field proportional to $qs/d^3$, and for the induced dipole $s$ is proportional to $1/d^2$.

**RQ3-8:**

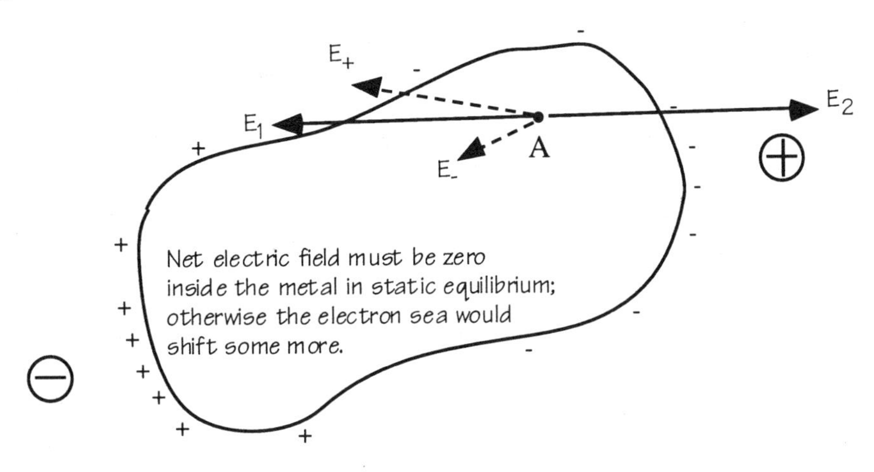

**HW3-5:**

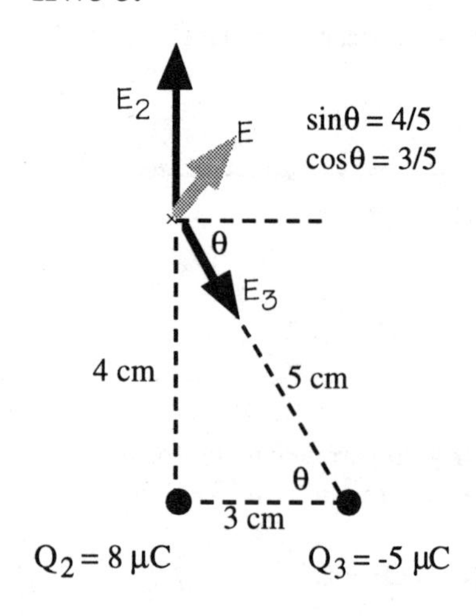


(a) Remove $Q_1$, draw electric fields due to $Q_2$ and $Q_3$. calculate magnitudes of $E_2$ and $E_3$:

$$E_2 = \left(9\times10^9 N-m^2/C^2\right)\frac{8\times10^{-6}C}{(.04m)^2} = 4.5\times10^7\frac{N}{C}$$

$$E_3 = \left(9\times10^9 N-m^2/C^2\right)\frac{5\times10^{-6}C}{(.05m)^2} = 1.8\times10^7\frac{N}{C}$$

Next calculate the components of the net field:

$E_x = E_{2x} + E_{3x} = E_3\cos\theta = (1.8\times10^7\ N/C)(3/5)$

$= 1.08\times10^7\ N/C$

$E_y = E_{2y} + E_{3y} = E_2 - E_3\sin\theta$

$= 4.5\times10^7\ N/C - (1.8\times10^7\ N/C)(4/5)$

$= 3.06\times10^7\ N/C$

If we put $Q_1$ at that location, it will experience a force $\vec{F}_{on\,Q_1} = Q_1\vec{E}$:

$$F_x = Q_1E_x = \left(3\times10^{-6}C\right)\left(1.08\times10^7 N/C\right) = 32.4N$$

$$F_y = Q_1E_y = \left(3\times10^{-6}C\right)\left(3.06\times10^7 N/C\right) = 91.8N$$

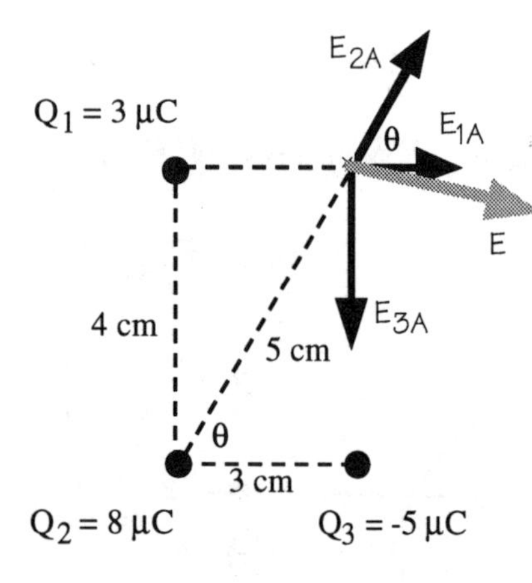


(b) Draw electric fields due to $Q_1$, $Q_2$, and $Q_3$. calculate magnitudes of $E_{1A}$, $E_{2A}$, and $E_{3A}$:

$$E_{1A} = \left(9\times10^9 N-m^2/C^2\right)\frac{3\times10^{-6}C}{(.03m)^2} = 3.00\times10^7\frac{N}{C}$$

$$E_{2A} = \left(9\times10^9 N-m^2/C^2\right)\frac{8\times10^{-6}C}{(.05m)^2} = 2.88\times10^7\frac{N}{C}$$

$$E_{3A} = \left(9\times10^9 N-m^2/C^2\right)\frac{5\times10^{-6}C}{(.04m)^2} = 2.81\times10^7\frac{N}{C}$$

Next calculate the components of the net field:

$E_{Ax} = E_{1Ax} + E_{2Ax} + E_{3Ax} = E_{1A} + E_{2A}\cos\theta$

$= (3.00\times10^7\ N/C) + (2.88\times10^7\ N/C)(3/5)$

$= 4.73\times10^7\ N/C$

$E_{Ay} = E_{1Ay} + E_{2Ay} + E_{3Ay} = E_{2A}\sin\theta - E_{3A}$

$= (2.88\times10^7\ N/C)(4/5) - (2.81\times10^7\ N/C)$

$= -0.51\times10^7\ N/C$

(c) From Newton's second law,

$$\vec{a} = \frac{q\vec{E}}{m} = \frac{2 \times 1.6 \times 10^{-19}\,C}{\left(\dfrac{4 \times 10^{-3}}{6 \times 10^{23}}\,kg\right)}\vec{E} = \left(4.8 \times 10^{7}\,\frac{C}{kg}\right)\vec{E}$$

We already calculated the components of the electric field, so

$$a_x = \left(4.8 \times 10^{7}\,\frac{C}{kg}\right)\left(4.73 \times 10^{7}\,\frac{N}{C}\right) = 2.27 \times 10^{15}\,\frac{m}{s^2}$$

$$a_y = \left(4.8 \times 10^{7}\,\frac{C}{kg}\right)\left(-0.51 \times 10^{7}\,\frac{N}{C}\right) = -2.45 \times 10^{14}\,\frac{m}{s^2}$$

**HW3-6:** By charge conservation, the charge on the silk must be -5nC.

(a) The arrows show the approximate direction and magnitude of the electric field at the various locations, using dashed lines for the contributions of the silk and glass, and drawing the final electric field vectors with solid lines. Some of the small vectors should really be drawn even smaller ($1/d^2$ falls off FAST), but they would then be hard to see on this diagram.

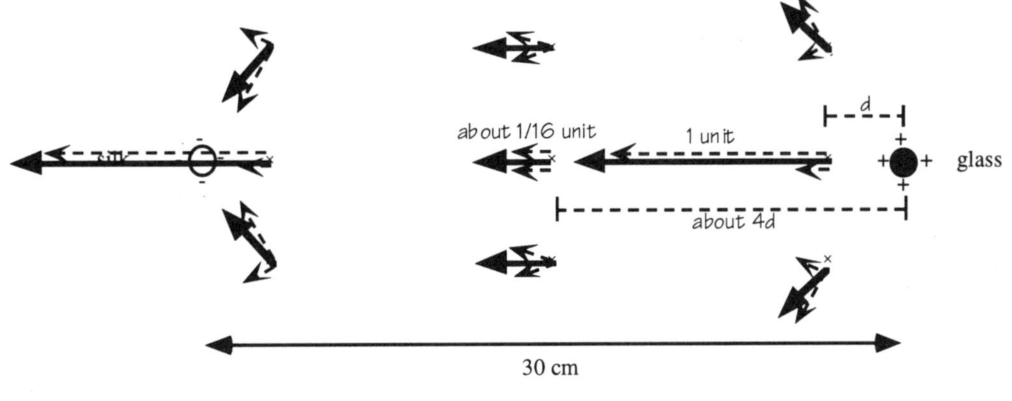


(b)

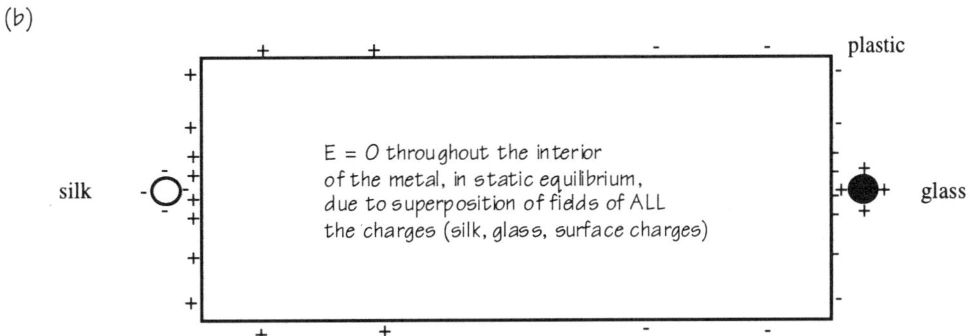


(c) The polarization of a molecule depends on the electric field at that location. Given the net field shown in part (a), the polarization of molecules at these locations would look like this:

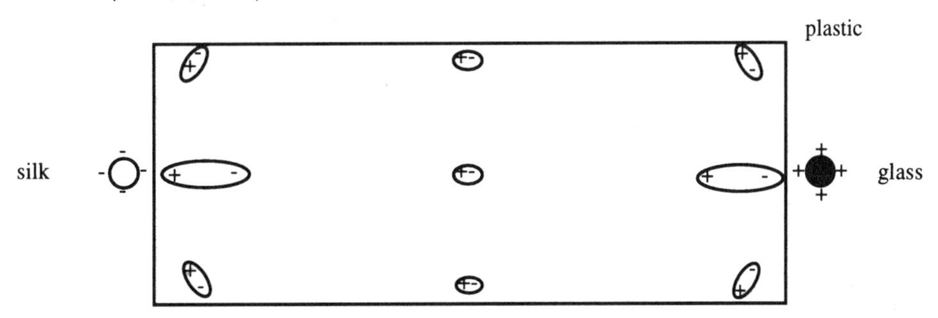

**(d)** With the plastic in place, the force on the glass ball is *greater* than the attraction by the silk, because the polarized molecules *also* attract the ball:

$$F_{\text{on glass ball}} > \left(9\times10^{9}\,\text{N}-\text{m}^2/\text{C}^2\right)\frac{\left(5\times10^{-9}\,\text{C}\right)^2}{(.3\text{m})^2} = 2.5\cdot10^{-6}\ \text{N}$$

**HW3-8:**

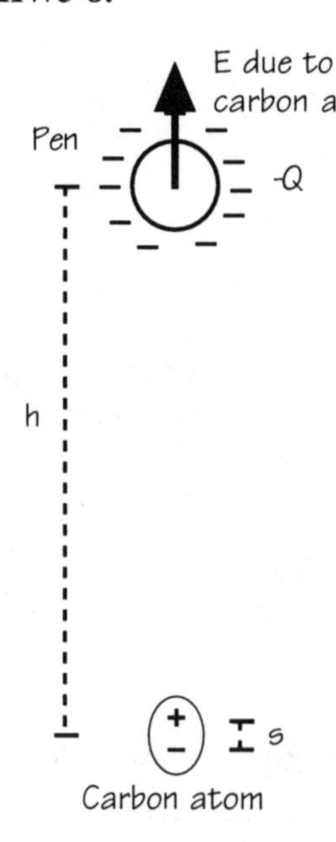


(a) The field made by the polarized carbon atom, at the location of the pen, is $E_{carbon} = \dfrac{1}{4\pi\varepsilon_0}\dfrac{2s(4e)}{h^3}$

The force exerted by the carbon atom on the pen is

$$F_{\substack{\text{carbon}\\\text{on pen}}} = \left|Q_{pen}\right|E_{carbon} = \frac{1}{4\pi\varepsilon_0}\frac{2s(4e)Q}{h^3}$$

(b) By Newton's Third Law, the force exerted by the pen on the carbon atom is the same in magnitude as the force exerted by the carbon atom on the pen:

$$F_{\substack{\text{pen on}\\\text{carbon}}} = \frac{1}{4\pi\varepsilon_0}\frac{2s(4e)Q}{h^3}$$

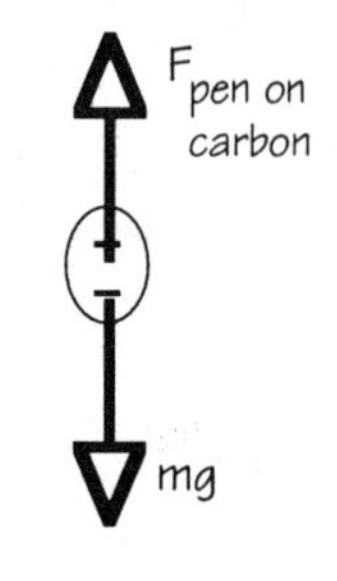


(c) The paper is lifted when the force by the pen on each carbon atom is equal to the weight of the atom:

$$\frac{1}{4\pi\varepsilon_0}\frac{2s(4e)Q}{h^3} = mg$$

$$s = \frac{4\pi\varepsilon_0 h^3 mg}{8eQ}$$

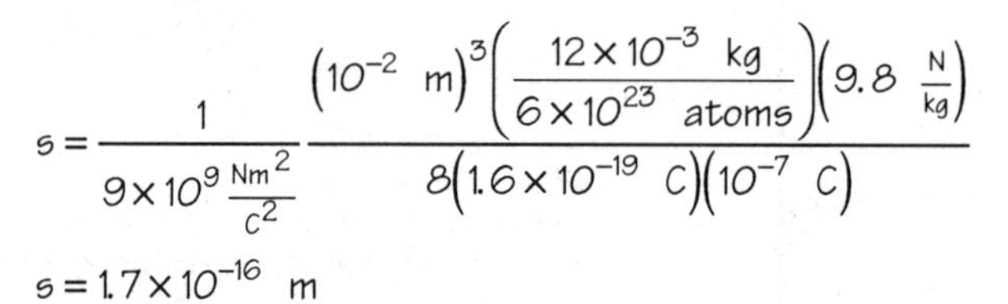

$$s = \frac{1}{9\times10^{9}\,\frac{\text{Nm}^2}{\text{C}^2}}\;\frac{\left(10^{-2}\ \text{m}\right)^3\left(\dfrac{12\times10^{-3}\ \text{kg}}{6\times10^{23}\ \text{atoms}}\right)\left(9.8\ \frac{\text{N}}{\text{kg}}\right)}{8\left(1.6\times10^{-19}\ \text{C}\right)\left(10^{-7}\ \text{C}\right)}$$

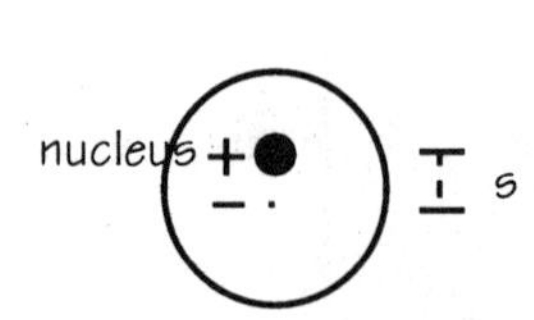


$$s = 1.7\times10^{-16}\ \text{m}$$

The diameter of the atom is about $10^{-10}$ m, so s is very small compared to the diameter of the atom.

(d) $p = (4e)s = \alpha E_{pen}$, where $E_{pen}$ is the field made by the pen, which polarizes the carbon atom:

$$\alpha = \frac{4es}{E_{pen}} = \frac{(4e)\left(\dfrac{4\pi\varepsilon_0 h^3 mg}{8eQ}\right)}{\dfrac{1}{4\pi\varepsilon_0}\left(\dfrac{Q}{h^2}\right)} = \left(4\pi\varepsilon_0\right)^2\frac{mgh^5}{2Q^2} = \left(\frac{1}{9\times10^{9}\,\frac{\text{N}-\text{m}^2}{\text{C}^2}}\right)^2\frac{\left(2\times10^{-26}\text{kg}\right)\left(9.8\frac{\text{N}}{\text{kg}}\right)\left(10^{-2}\text{m}\right)^5}{2\left(10^{-7}\text{C}\right)^2}$$

$$\alpha = 1.2\times10^{-41}\,\frac{\text{m}^3}{\frac{\text{N}-\text{m}^2}{\text{C}^2}}$$

This is fairly good agreement, since $\alpha$ depends on $h^5$, and we only estimated h.

# Chapter 4 solutions

**HW4-2:**

(a)

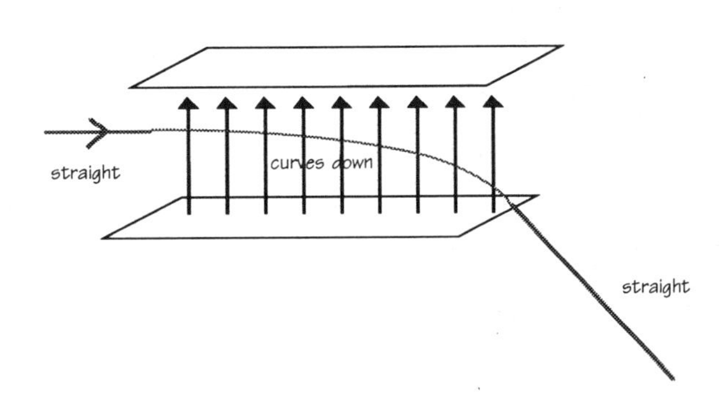


(b)

$$ma = F = |qE|$$
$$a = \frac{F}{m} = \frac{|qE|}{m}$$
$$= \frac{(1.6\text{x}10^{-19}C)(10^5 N/C)}{9\text{x}10^{-31}kg}$$
$$= 1.8\text{x}10^{16} ms^{-2}$$

The force on the electron due to the electric field of the plates.

This is a huge acceleration, but it is reasonable because the mass of the electron is so small.

(c)

$$E = \frac{Q/A}{\varepsilon_0}$$
$$Q = E\varepsilon_0 A$$
$$= (10^5 N/C)(9\text{x}10^{-12} C^2N^{-1}m^{-2})(.12m\text{x}.03m)$$
$$= 3.2\text{x}10^{-9} C$$

Field between two large plates.

The upper plate is negative, because the electric field vectors point toward it.

(d) The field just outside a capacitor is reduced by approximately $\frac{d}{L}$, where L is the length or width of the plate, and d is the gap distance (page 154). So the field outside a plate is approximately:

$$E \approx (10^5 N/C)\left(\frac{.0025m}{.03m}\right)$$
$$\approx 8\text{x}10^3 N/C$$

**HW4-3:**

(a)

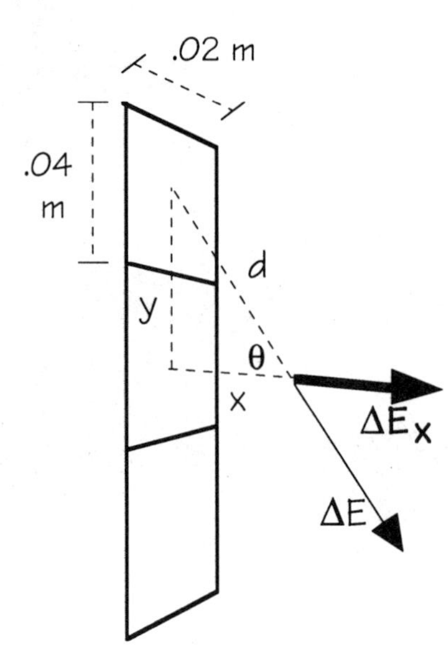


1. Cut up distribution into pieces and draw ΔE. Problem statement specifies 3 pieces, each 2 cm wide by 4 cm high. ΔE for top piece is shown in diagram. It is clear from the diagram that we only need the x component, since the y components will cancel.

2. Algebraic expression for ΔE for each piece: Consider each piece as a point charge, located at center of rectangle. Then:

$$\Delta E = \frac{1}{4\pi\varepsilon_0}\frac{\Delta Q}{d^2}, \text{ and we want x-component:}$$

$$\Delta E_x = \frac{1}{4\pi\varepsilon_0}\frac{\Delta Q}{d^2}\cos\theta$$

$$d = \sqrt{x^2+y^2}, \text{ and } \cos\theta = \frac{x}{d} = \frac{x}{\sqrt{x^2+y^2}}, \text{ and } \Delta Q = \left(Q/A\right)\Delta A$$

(where $Q$ = total charge on tape,
$A$ = total area, $\Delta A = \Delta y \Delta z$ = area of small piece.)

$$\Delta E_x = \frac{1}{4\pi\varepsilon_0}\frac{\left(Q/A\right)\Delta A}{x^2+y^2}\frac{x}{\sqrt{x^2+y^2}} = \frac{1}{4\pi\varepsilon_0}\left(Q/A\right)\Delta A\frac{x}{\left(x^2+y^2\right)^{\frac{3}{2}}}$$

4. Add up all contributions:

Since we are at point in middle, y and z components of E will add up to zero, and $E_{net}$ will be in x direction. We only need to find x-component of ΔE for each piece. $E_x = \sum \Delta E_x = \frac{1}{4\pi\varepsilon_0}\left(Q/A\right)\Delta A \sum_{i=1}^{3}\frac{x_i}{\left(x_i^2+y_i^2\right)^{\frac{3}{2}}}$

For this calculation,

$$\frac{1}{4\pi\varepsilon_0}\left(Q/A\right)\Delta x\Delta y = \left(9\times10^9\,\frac{\text{N-m}^2}{\text{C}^2}\right)\left(\frac{4\times10^{-8}\text{ C}}{(.02\text{m})(.12\text{m})}\right)(.02\text{m})(.04\text{m})$$

$$= 120\,\frac{\text{N-m}^2}{\text{C}}$$

Only the last term depends on distance, so we'll calculate it for each piece:

| Piece | x | y | z | $x/(x^2+y^2)^{1.5}$ |
|---|---|---|---|---|
| 1 | .03 m | .04 m | 0 | 240 |
| 2 | .03m | 0.00 m | 0 | 1111 |
| 3 | .03 m | .04 m | 0 | 240 |
| | | | Total: | 1591 |

$$\text{So } E_x = \left(120\,\frac{\text{N-m}^2}{\text{C}}\right)\left(1591\text{ m}^{-2}\right) = 1.91\times10^5\,\frac{\text{N}}{\text{C}}$$

4. Check that result is reasonable: (a) The units come out right (N/C). (b) This is less than the critical field for breakdown of air ($3\times10^6$ N/C), so it is a possible value. (c) Also, a point charge of $4\times10^{-8}$ C located 3 cm away would make a field of $(9\times10^9)(4\times10^{-8})/(.01)^2 = 4\times10^5$ N/C, and our E is smaller than that, as it should be, since almost all of the tape is farther away than 3 cm..

(b)

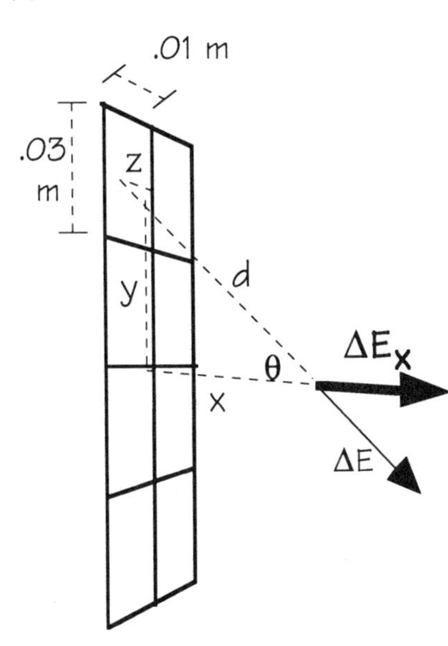


1. Cut up distribution into pieces and draw $\Delta E$. Problem statement specifies 8 pieces, each 1 cm wide by 3 cm high. $\Delta E$ is shown.

2. Algebraic expression for $\Delta E$ for each piece: Again consider each piece as a point charge, located at center of rectangle. Then:

$\Delta E = \frac{1}{4\pi\varepsilon_0}\frac{\Delta Q}{d^2}$, and we want only x-component (others cancel):

$$\Delta E_x = \frac{1}{4\pi\varepsilon_0}\frac{\Delta Q}{d^2}\cos\theta$$

$d = \sqrt{x^2+y^2+z^2}$, and $\cos\theta = \frac{x}{d} = \frac{x}{\sqrt{x^2+y^2+z^2}}$, and $\Delta Q = \left(Q/A\right)\Delta A$

(where $Q$ = total charge on tape,
$A$ = total area, $\Delta A = \Delta y \Delta z$ = area of small piece.)

$$\Delta E_x = \frac{1}{4\pi\varepsilon_0}\frac{\left(Q/A\right)\Delta A}{x^2+y^2+z^2}\frac{x}{\sqrt{x^2+y^2+z^2}} = \frac{1}{4\pi\varepsilon_0}\left(Q/A\right)\Delta A\frac{x}{\left(x^2+y^2+z^2\right)^{\frac{3}{2}}}$$

3. Add up all the contributions: $E_x = \sum \Delta E_x = \frac{1}{4\pi\varepsilon_0}\left(Q/A\right)\Delta A \sum_{i=1}^{8}\frac{x_i}{\left(x_i^2+y_i^2+z_i^2\right)^{\frac{3}{2}}}$

This time the pieces are smaller, so:

$$\frac{1}{4\pi\varepsilon_0}\left(Q/A\right)\Delta A = \left(9\times10^9\ \frac{\text{N-m}^2}{\text{C}^2}\right)\left(\frac{4\times10^{-8}\ \text{C}}{(.02\text{m})(.12\text{m})}\right)(.01\text{m})(.03\text{m})$$
$$= 45\ \frac{\text{N-m}^2}{\text{C}}$$

As before, only the part inside the summation varies with distance, so we'll calculate it for each piece. Notice that there are actually only two different calculations to do, because of symmetry:

| Piece | x | y | z | $x/(x^2+y^2+z^2)^{1.5}$ |
|---|---|---|---|---|
| 1 | .03 m | .045 m | .005 m | 187.2 |
| 2 | .03 m | .045 m | .005 m | 187.2 |
| 3 | .03 m | .015 m | .005 m | 769.3 |
| 4 | .03 m | .015 m | .005 m | 769.3 |
| 5 | .03 m | .015 m | .005 m | 769.3 |
| 6 | .03 m | .015 m | .005 m | 769.3 |
| 7 | .03 m | .045 m | .005 m | 187.2 |
| 8 | .03 m | .045 m | .005 m | 187.2 |
| | | | Total: | 3826 |

$$\text{So } E_x = \left(45\frac{\text{N-m}^2}{\text{C}}\right)\left(3825\ \text{m}^{-2}\right) = 1.72\times10^5\ \frac{\text{N}}{\text{C}}$$

4. <u>Check that result is reasonable:</u> This is less than the critical field for breakdown of air ($3\times10^6$ N/C), so it is a possible value. The units come out right (N/C). As in (a), this is less than the field of a point charge 3 cm away.

4-3 c: In general, the more pieces the distribution is divided into, the more accurate the calculation will be. So the answer from (b) is closer to the actual electric field than the answer to (a). In addition, (a) is probably much too high because taking one large piece at y=0 significantly underestimates the average distance to the charges.

We wrote a computer program to calculate $E_x$ for a number of different ways of cutting up the tape. The table gives the results. Here $n_y$ is the number of pieces in the y direction (height), and $n_z$ is the number of pieces in the z direction (width).

| $n_y$ | $n_z$ | $E_x$ (N/C) |
|---|---|---|
| 3 | 1 | $1.909\times10^5$ (part a) |
| 4 | 2 | $1.722\times10^5$ (part b) |
| 8 | 4 | $1.726\times10^5$ |
| 20 | 4 | $1.725\times10^5$ |
| 100 | 10 | $1.721\times10^5$ |
| 200 | 20 | $1.721\times10^5$ |

Eventually we do not gain much by adding more pieces.

**HW4-4:**

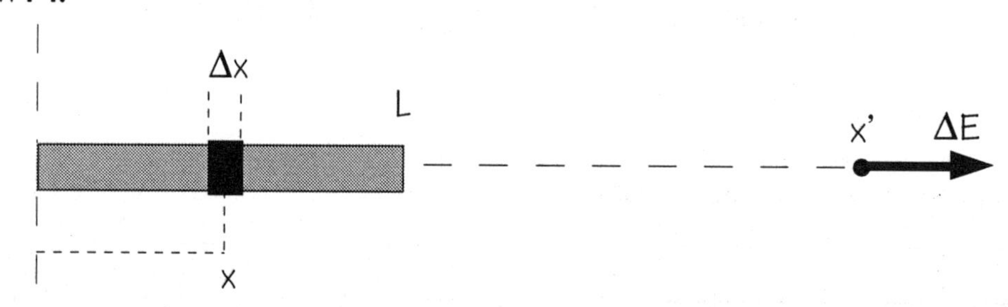


1) Cut up the distribution into point-like slices, of width Δx. ΔE for one piece is shown.

2) To write an expression for ΔE, we define a variable x that is the distance from the origin to the center of a slice of the rod. Then the distance from the slice to point x' is (x'-x).

   The amount of charge Δq on a slice of length Δx is given by the charge per unit length, Q/L, times the length of the slice: $\Delta q = \frac{Q}{L}\Delta x$. So

$$\Delta E = \frac{1}{4\pi\varepsilon_0}\frac{\Delta q}{(x'-x)^2} = \frac{1}{4\pi\varepsilon_0}\frac{Q}{L}\frac{\Delta x}{(x'-x)^2}$$

3) Adding up the contributions of all the slices means integrating, with x running from 0 to L:

$$E = \int_0^L \frac{1}{4\pi\varepsilon_0}\frac{Q}{L}\frac{dx}{(x'-x)^2} = \frac{1}{4\pi\varepsilon_0}\frac{Q}{L}\int_0^L \frac{dx}{(x'-x)^2} = \frac{1}{4\pi\varepsilon_0}\frac{Q}{L}\frac{1}{(x'-x)}\Bigg|_0^L = \frac{1}{4\pi\varepsilon_0}\frac{Q}{L}\left[\frac{1}{(x'-L)} - \frac{1}{x'}\right]$$

$$E = \frac{1}{4\pi\varepsilon_0}\frac{Q}{x'(x'-L)}$$

5) Check: a) units: $\left(\frac{Nm^2}{C^2}\right)\frac{C}{m \cdot m} = \frac{N}{C}$ which is correct for electric field

b) if x >> L, $E \to \frac{1}{4\pi\varepsilon_0}\frac{Q}{x'^2}$, and the rod looks like a point charge, as it should.

**HW4-7:**

(a) Location 1 is inside the uniformly charged sphere, so the sphere does not contribute to the net field there; the only nonzero contribution is from the plates. Since s is very small, in this region $E_{plate} \approx \frac{Q_{plate}/A_{plate}}{2\varepsilon_0}$, so $|\vec{E}_1| = |\vec{E}_+| + |\vec{E}_-| = \frac{Q/A}{2\varepsilon_0} + \frac{2Q/A}{2\varepsilon_0} = \frac{3Q/A}{2\varepsilon_0}$ to the right

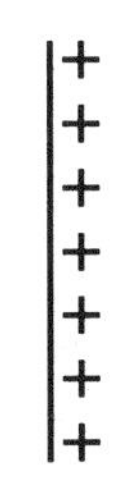
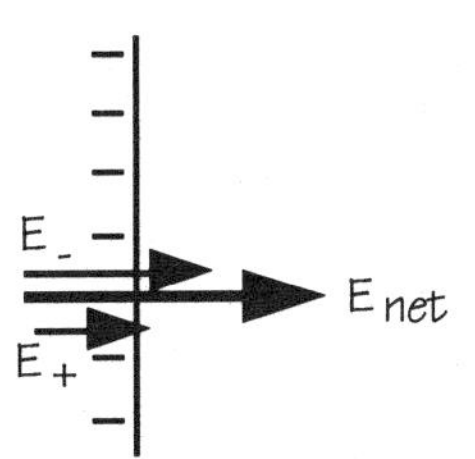


b) At location 2, the sphere also contributes.
Net field to the right is:

$$E_2 = E_1 - \frac{1}{4\pi\varepsilon_0}\frac{2Q}{(3r)^2}$$
$$= \frac{3Q/A}{\varepsilon_0} - \frac{1}{4\pi\varepsilon_0}\frac{2Q}{9r^2}$$

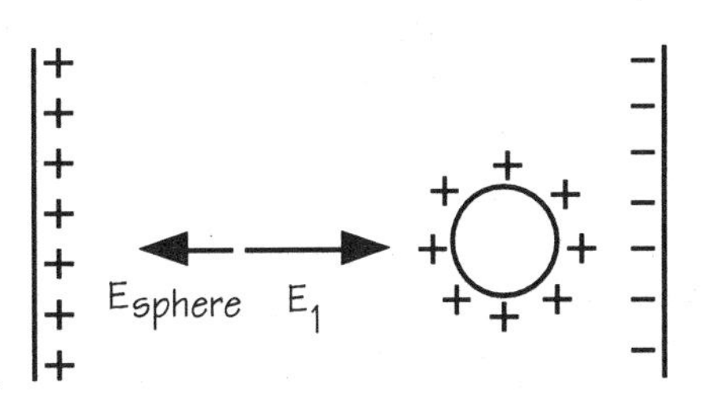


If this turns out to be negative, then the net field is to the left.

(c) At location 3, we have:

$$E_{1x} = E_1$$
$$E_{1y} = 0$$
$$E_{sphere} = \frac{1}{4\pi\varepsilon_0}\frac{2Q}{\left(3\sqrt{2}r\right)^2}$$

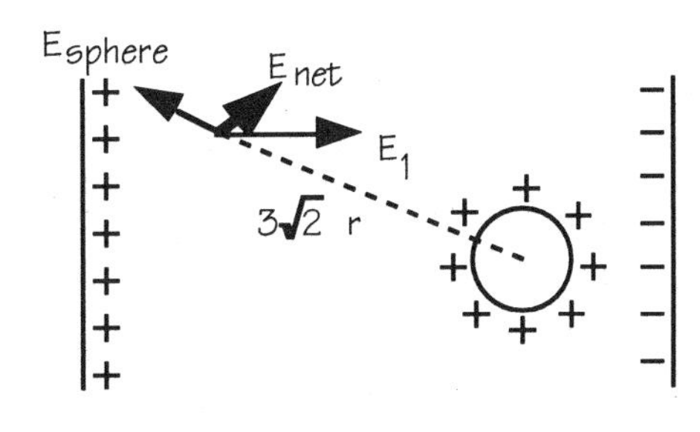


$$E_{3x} = E_1 - E_{sphere}\cos 45^\circ$$
$$= \frac{3Q/A}{2\varepsilon_0} - \frac{1}{4\pi\varepsilon_0}\frac{Q}{9r^2}\frac{\sqrt{2}}{2}$$
$$E_{3y} = 0 + E_{sphere}\sin 45^\circ$$
$$= +\frac{1}{4\pi\varepsilon_0}\frac{Q}{9r^2}\frac{\sqrt{2}}{2}$$

**HW4-9:**

(a) E is largest near the surface of the sphere, so if the field exceeds the breakdown limit anywhere, it will certainly be above the limit near the surface.

(b& c) The critical charge Q is that charge that makes the field very near the surface equal the critical field:.

$$E = \frac{1}{4\pi\varepsilon_0}\frac{Q}{R^2} = 3\times 10^6\ \frac{N}{C}$$

$$Q = \frac{\left(3\times 10^6\ \frac{N}{C}\right)R^2}{\left(9\times 10^9\ \frac{N\ m^2}{C^2}\right)}$$

$$R = 0.1\ m \Rightarrow Q = 3.3\times 10^{-6}\ C$$

$$R = 0.001\ m \Rightarrow Q = 3.3\times 10^{-10}\ C$$

(d) All through the shaded volume surrounding the sphere (exaggerated in the diagram), the electric field is greater than the critical field, and the air ionizes. Electrons are attracted to the metal sphere and reduce its positive charge. Positive ions move away from the sphere and end up on the walls, ceiling, and floor.

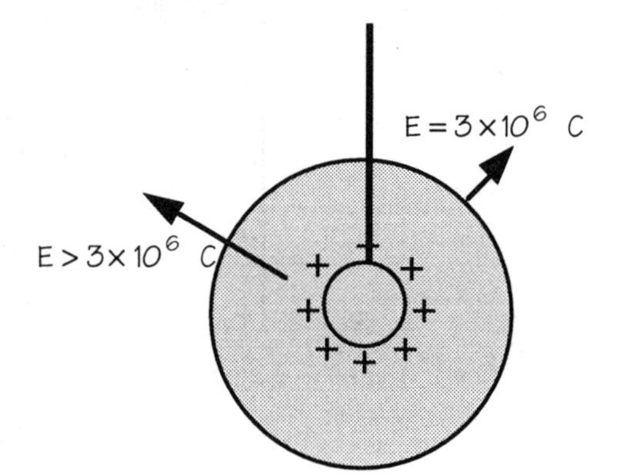


The volume of ionized air contracts as the charge on the sphere decreases. When the positive charge on the sphere is reduced to the point that $E < 3\times 10^6\ \frac{N}{C}$ near the surface, the process stops. The final charge on the ball plus charge on the walls, ceiling, and floor is equal to the original charge on the ball (charge conservation).

**HW4-12:**

(a)

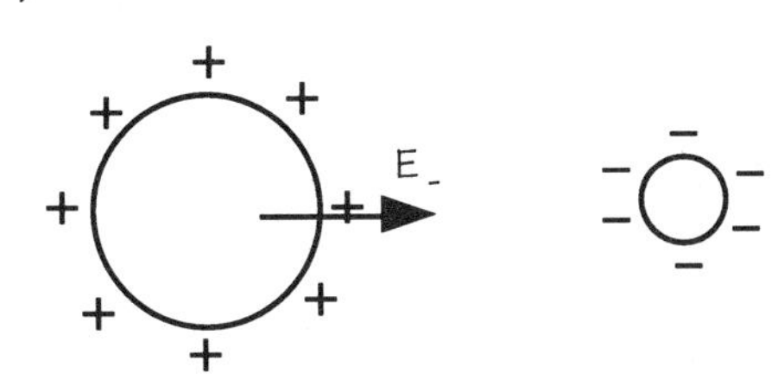


At A, we are inside the positive sphere, so only the negative sphere contributes:

$$E_{Ax} = \left(9\times10^9 \frac{\text{N m}^2}{\text{C}^2}\right)\frac{\left(10^{-9}\text{C}\right)}{(.19\text{m})^2} = 249\frac{\text{N}}{\text{C}}$$

$$E_{Ay} = 0$$

(b)

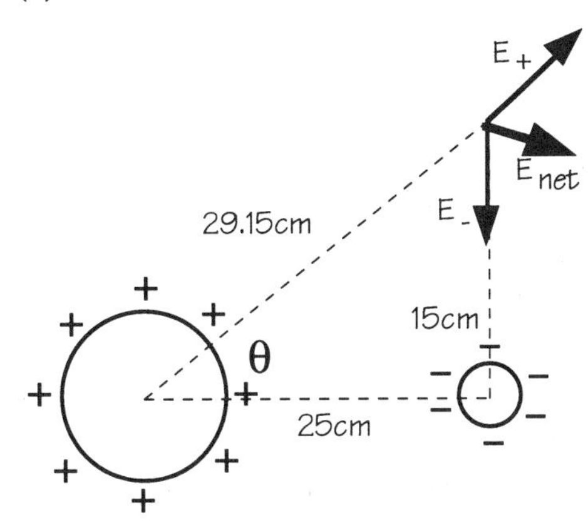


At B, both spheres contribute. Outside the spheres they look like point charges:

$$E_+ = \left(9\times10^9 \ \frac{\text{Nm}^2}{\text{C}^2}\right)\frac{\left(4\times10^{-9}\ \text{C}\right)}{(.2915\text{m})^2} = 424\ \frac{\text{N}}{\text{C}}$$

$$E_- = \left(9\times10^9 \ \frac{\text{Nm}^2}{\text{C}^2}\right)\frac{\left(1\times10^{-9}\ \text{C}\right)}{(.15\text{m})^2} = 400\ \frac{\text{N}}{\text{C}}$$

$$\tan\theta = \frac{15}{25};\quad \sin\theta = \frac{15}{29.15};\quad \cos\theta = \frac{25}{29.15}$$

$$E_{Bx} = E_+\cos\theta = 424\left(\frac{25}{29.15}\right)\frac{\text{N}}{\text{C}} = 364\ \frac{\text{N}}{\text{C}}$$

$$E_{By} = E_+\sin\theta - E_- = \left[424\left(\frac{15}{29.15}\right) - 400\right]\ \frac{\text{N}}{\text{C}} = -182\ \frac{\text{N}}{\text{C}}$$

(c)

$$F_{Bx} = -eE_{Bx} = \left(-1.6\times10^{-19}\ \text{C}\right)\left(364\ \tfrac{\text{N}}{\text{C}}\right) = -5.82\times10^{-17}\ \text{N}$$

$$F_{By} = -eE_{By} = \left(-1.6\times10^{-19}\ \text{C}\right)\left(-182\ \tfrac{\text{N}}{\text{C}}\right) = +2.91\times10^{-17}\ \text{N}$$

**HW4-13:**

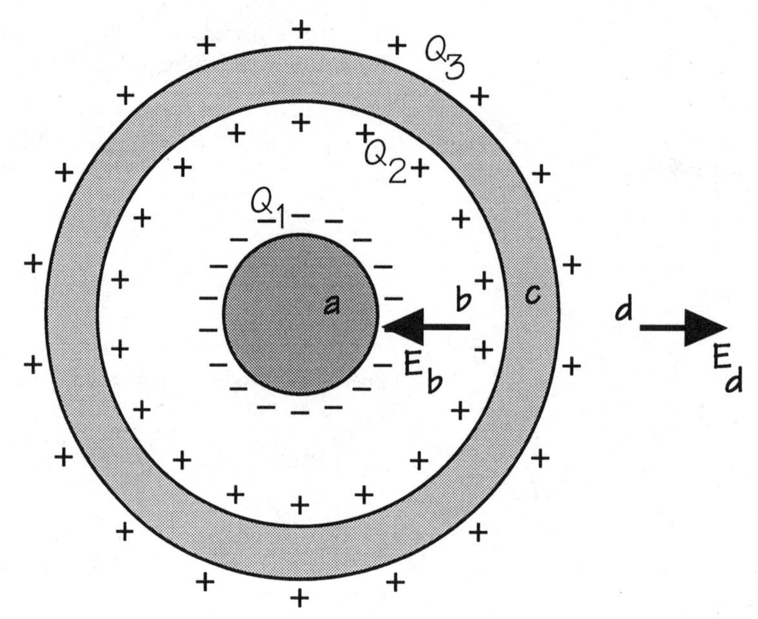


There are 3 concentric uniform spherical shells of charge here: $Q_1$ (on the surface of the plastic sphere), $Q_2$ (on the inner surface of the metal shell), and $Q_3$ (on the outer surface of the metal shell). At any point, the net electric field is the sum of the contributions from all 3 of these uniform spheres of charge.

(a) Point a is inside the plastic sphere, so it is <u>inside</u> all of the spherical shells of charge. Since inside a uniform spherical charge distribution, E due to that sphere =0, each sphere contributes nothing to the electric field at a. $E_a=0$.

(b) Point b is in the air gap. It is <u>inside</u> spheres 2 and 3, so $Q_2$, and $Q_3$, contribute nothing to $E_b$. However it is <u>outside</u> sphere 1, so sphere 1 looks like a point charge located at the center of the sphere. So

$$E_b = \frac{1}{4\pi\varepsilon_0}\frac{Q_1}{r^2}, \text{ radially inward}$$

(c) Point c is <u>inside</u> the metal sphere. Since in static equilibrium, E=0 inside a metal, $E_c=0$.

(d) Point d is <u>outside</u> all 3 spheres of charge. Thus all three contribute to E, and:

$$E_d = \frac{1}{4\pi\varepsilon_0}\frac{Q_3}{r^2} + \frac{1}{4\pi\varepsilon_0}\frac{Q_2}{r^2} - \frac{1}{4\pi\varepsilon_0}\frac{Q_1}{r^2} = \frac{1}{4\pi\varepsilon_0}\frac{1}{r^2}(Q_3 + Q_2 - Q_1), \text{ radially out}$$

(e) Consider point c. The net electric field there is zero, but it must be the sum of the contributions of charge distributions $Q_1$ and $Q_2$, since this point is outside both spherical shells of charge. Hence at point c:

$$\vec{E}_1 + \vec{E}_2 = 0$$

$$\frac{1}{4\pi\varepsilon_0}\frac{Q_1}{r^2} = \frac{1}{4\pi\varepsilon_0}\frac{Q_2}{r^2}$$

$$Q_1 = Q_2$$

So Q2 = +5nC

(f) Inside the plastic sphere the net electric field = 0, since every point inside the sphere is <u>inside</u> each of the three spherical shells of charge. Since E=0, there is no force on the + and - charges within a molecule, so the molecules do not polarize. There is no polarization of molecules in the plastic sphere.

# Chapter 5 solutions

**HW5-1:**

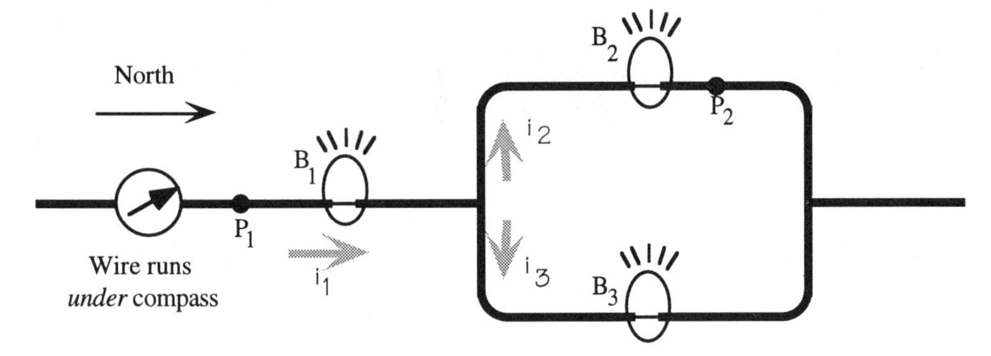


(a) Compass is on top of wire. $B_{wire}$ points up (west), so electron current must run to the right (north).

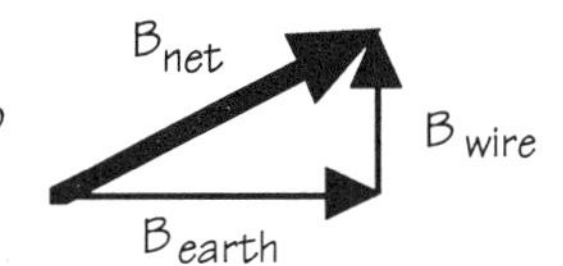


(b)

$3 \times 10^{18}$ electrons/s pass point P1. By charge conservation and definition of steady state, $i_1 = i_2 + i_3$. If bulbs 2 and 3 are identical,

$$i_2 = i_3 = \frac{i_1}{2} = \frac{3 \times 10^6 \ \text{electrons/s}}{2} = 1.5 \times 10^6 \ \text{electrons/s passing point } P_2.$$

(c) $B_1 > B_2 = B_3$ Twice as much current through $B_1$.

(d)

In tungsten there are $6.3 \times 10^{28} \ \frac{\text{free electrons}}{m^3}$

Cross-sectional area of $B_1 = 10^{-8} m^2$

Electron mobility in hot tungsten is $1.2 \times 10^{-4} \ \frac{m/s}{N/C}$

$i_1 = nAv_1 = nAuE_1$

$$E_1 = \frac{i_1}{nAu} = \frac{3 \times 10^{18} \ \text{electrons/s}}{\left(6.3 \times 10^{28} \ \frac{\text{electrons}}{m^3}\right)\left(10^{-8} m^2\right)\left(1.2 \times 10^{-4} \ \frac{m/s}{N/C}\right)} = 39.7 \ \frac{N}{C} \text{ to the left}$$

(direction of E is opposite to direction of i).

## Chapter 6 solutions

**RQ6-4:**

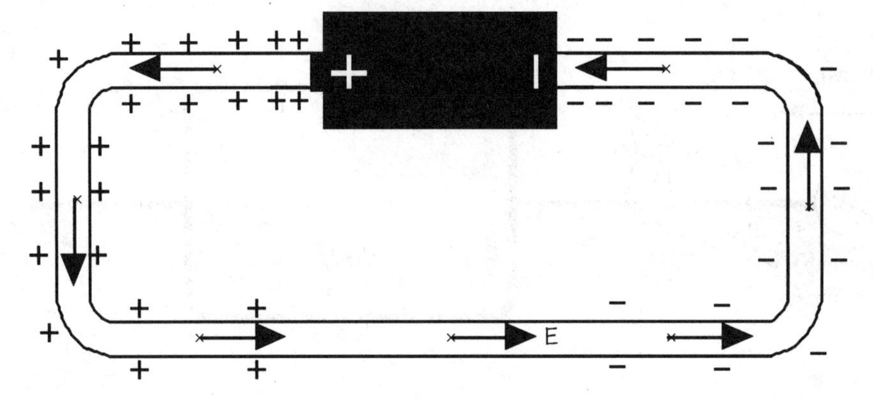


There is a fairly smooth gradient of surface charge all around the circuit, producing a uniform electric field inside the wire.

**RQ6-5:**

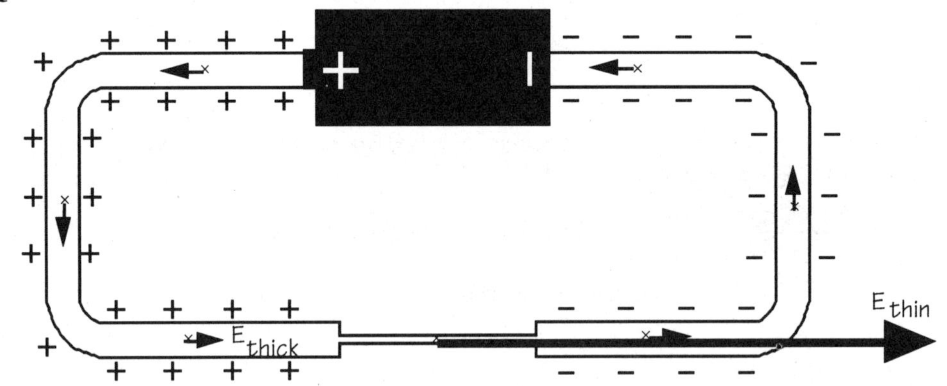


On the thick wires there is a smaller surface-charge gradient than in RQ6-4, making a smaller electric field inside the thick wires than in 6-4. There is a large gradient across the thin wire, making a much bigger electric field inside the thin wire. This large electric field is needed to produce a high drift speed for electrons inside the thin wire, compensating for the difference in cross sectional areas, so the current in the thin and thick wires can be the same. By comparing the electric field in the thick wires in 6-5 and 6-4, we can see that the current in 6-5 is <u>less</u> than the current in 6-4 (smaller E, same n, A, and u in both cases).

**RQ6-7:**

| Experiment | Effect on current | Parameter(s) that changed |
|---|---|---|
| Double the length of a nichrome wire | $\frac{1}{2}i$ | n A u ***E*** $E = \frac{F_{NC}s}{e(2L)}$ |
| Double the cross-sectional area of a nichrome wire | 2i | n ***A*** u E Ax2 |
| Two identical bulbs in series compared to a single bulb | ~ 0.7i | n A ***u*** ***E*** $E = \frac{F_{NC}s}{e(2L)}$, u bigger |
| Two batteries in series compared to a single battery | 2i (for nichrome wire) | n A u ***E*** $E = \frac{2(F_{NC}s)}{eL}$ |

**RQ6-9:** answer: 8 degrees

**HW6-1:**

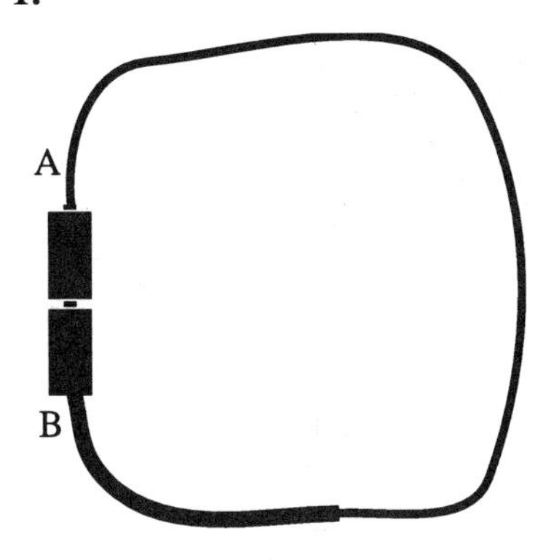


Thin wire:

$L_1 = .5\text{m}$

$$A_1 = \pi\left(\frac{.25\times10^{-3}\text{m}}{2}\right)^2$$

Thick wire:

$L_2 = .15\text{m}$

$$A_2 = \pi\left(\frac{.35\times10^{-3}\text{m}}{2}\right)^2$$

(a) Current conservation: At steady state, current in thick and thin wires must be equal:

$$nA_1uE_1 = nA_2uE_2 \quad \text{(n and u same for both wires)}$$

$$\left(\frac{A_1}{A_2}\right)E_1 = E_2$$

Energy conservation: In steady state energy input from batteries = heat dissipated in circuit

$$2\left(2.4\times10^{-19}\ \frac{\text{J}}{\text{electron}}\right) = eE_1L_1 + eE_2L_2 = eE_1\left(L_1 + \left(\frac{A_1}{A_2}\right)L_2\right)$$

Solving for $E_1$ and $E_2$:

$$E_1 = \frac{2\left(2.4\times10^{-19}\ \frac{\text{J}}{\text{electron}}\right)}{\left(1.6\times10^{-19}\ \frac{\text{C}}{\text{electron}}\right)\left(.5\text{m} + \left(\frac{\pi\left(\frac{.25\times10^{-3}\text{m}}{2}\right)^2}{\pi\left(\frac{.35\times10^{-3}\text{m}}{2}\right)^2}\right)(.15\text{m})\right)} = 5.20\frac{\text{J}}{\text{C}-\text{m}} = 5.20\frac{\text{N}}{\text{C}}$$

$$E_2 = \left(\frac{\pi\left(\frac{.25\times10^{-3}\text{m}}{2}\right)^2}{\pi\left(\frac{.35\times10^{-3}\text{m}}{2}\right)^2}\right)\left(5.2\frac{\text{N}}{\text{C}}\right) = 2.65\frac{\text{N}}{\text{C}}$$

(b)

$$\text{Thin wire:}\quad v_1 = uE_1 = \left(7\times10^{-5}\frac{\text{m/s}}{\text{N/C}}\right)\left(5.2\frac{\text{N}}{\text{C}}\right) = 3.64\times10^{-4}\ \frac{\text{m}}{\text{s}}$$

$$\text{Thick wire:}\quad v_2 = uE_2 = \left(7\times10^{-5}\frac{\text{m/s}}{\text{N/C}}\right)\left(2.65\frac{\text{N}}{\text{C}}\right) = 1.86\times10^{-4}\ \frac{\text{m}}{\text{s}}$$

$$\text{time} = \frac{L_1}{v_1} + \frac{L_2}{v_2} = \left(\frac{.5\text{m}}{3.64\times10^{-4}\ \frac{\text{m}}{\text{s}}}\right) + \left(\frac{.15\text{m}}{1.86\times10^{-4}\ \frac{\text{m}}{\text{s}}}\right) = 2180\text{s}\left(\frac{1\ \text{min}}{60\text{s}}\right) = 36.3\ \text{min}$$

(c) Electric field propagates at about 1 ft / ns, or .3 m /ns. The shift in the electron sea takes very little time because each electron moves a very small distance (~$10^{-16}$ m in HW3-8, so the only significant time is the time it takes the change in E to travel .5 m, or about 1.5 ns.

$$\text{(d)}\quad i_1 = nA_1v_1 = \left(9\times10^{28}\ \frac{\text{electrons}}{\text{m}^3}\right)\left(4.91\times10^{-8}\text{m}^2\right)\left(3.64\times10^{-4}\ \frac{\text{m}}{\text{s}}\right) = 1.61\times10^{18}\ \frac{\text{electrons}}{\text{s}}$$

In the steady state the current is the same throughout the entire circuit.

**HW6-2:**

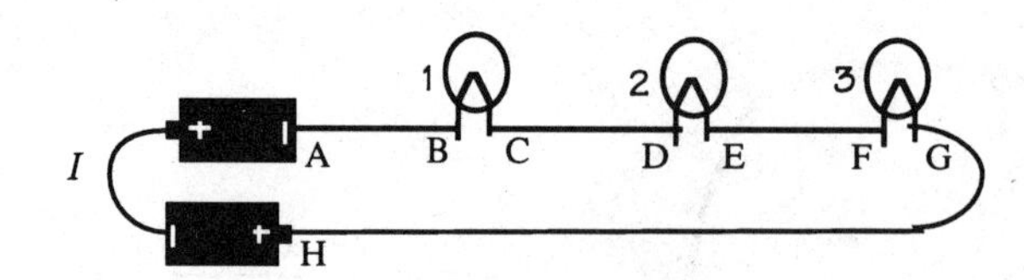


(a) In steady state current is the same everywhere in a series circuit, so $3 \times 10^{17}$ electrons/s enter D.

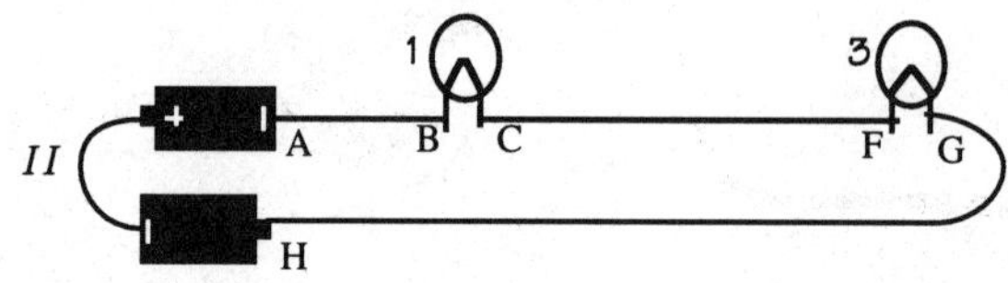


(b) The batteries in circuits I and II are the same, so energy input/electron is the same, and therefore energy dissipated by one electron in a round trip is the same. So, ignoring the wires, where the electric field is nearly zero:

Energy conservation: $W_{\substack{battery \\ per\ electron}} = F_{NC}s = 3E_I eL = 2E_{II}eL$

$$\frac{3}{2}E_I = E_{II}$$ There is a bigger electric field in the bulbs in circuit II than in circuit I

$$i_I = nAuE_I$$

$$i_{II} = nAuE_{II} = nAu\frac{3}{2}E_I = \frac{3}{2}i_I = \frac{3}{2}\left(3 \times 10^{17}\frac{\text{electrons}}{\text{s}}\right) = 4.5 \times 10^{17}\frac{\text{electrons}}{\text{s}}$$

(c) Two round bulbs → 22°. Three round bulbs →18°, so would predict $\frac{3}{2}(18°) = 27°$

This discrepancy is not surprising, because u decreases as T increases; as the bulbs in circuit II get hotter the mobility of their filaments decreases.

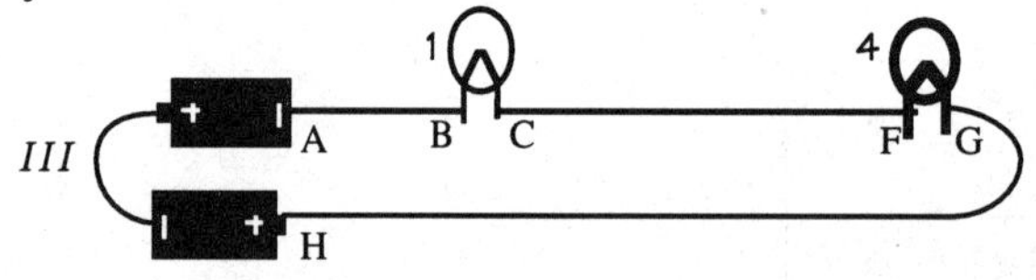


(d) Now $A_4 = 2A_1$

Current conservation:
$$nA_1uE_1 = nA_4uE_4 = n(2A_1)uE_4$$
$$E_4 = \frac{E_1}{2}$$

Energy conservation:
$$\text{Circuit } III\text{: } F_{NC}s = eE_1L + eE_4L = \frac{3}{2}eE_1L$$
$$\text{Circuit } I\text{: } F_{NC}s = 3eE_1^IL$$
$$3eE_1^IL = \frac{3}{2}eE_1L$$
$$2E_1^I = E_1$$

The electric field in bulb 1 is now twice as large as it was in circuit I. However, n, A, and u are still the same for bulb 1, so now the electron current through bulb 1 is:

$$i_{III} = nAu(2E_1^I) = (2)3 \times 10^{17}\frac{\text{electrons}}{\text{s}} = 6 \times 10^{17}\frac{\text{electrons}}{\text{s}}$$

**HW6-3:**

(a)

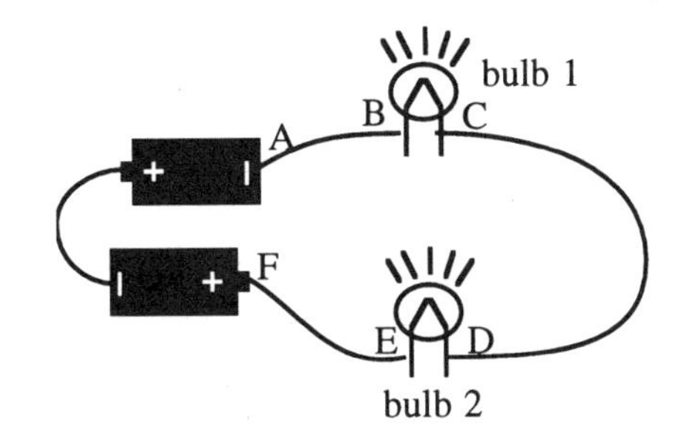


Current $i_1$ in bulb 1

Bulbs:

$A_1 = A_2 = A$

$L_1 = L_2 = L$

$n_1 = n_2 = n$

$3u_1 = u_2$

In the steady state $I_1 = I_2$ Current conservation

$$nAu_1E_1 = nAu_2E_2 = nA(3u_1)E_1$$

$$E_2 = \frac{E_1}{3}$$

(b)

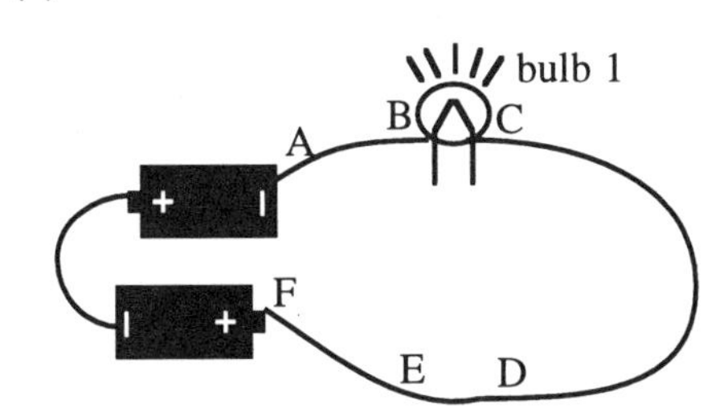


Current $i_O$,

Field $E_O$ in bulb

Drift speed $v_O$ in bulb

Energy conservation: energy input by batteries for one electron = work done on one electron as it goes around the whole circuit

$$\text{Circuit (a): } 2(F_{NC}s) = eE_1L + eE_2L + eE_{wires}L_{wires}$$

$$\approx eE_1L + eE_2L \quad (\text{because } E_{wires} \approx 0)$$

$$= \frac{4}{3}eE_1L$$

$$\text{Circuit (b): } 2(F_{NC}s) = eE_OL + eE_{wires}L_{wires}$$

$$\approx eE_OL \qquad (\text{because } E_{wires} \approx 0)$$

$$\text{Same batteries, so } eE_OL = \frac{4}{3}eE_1L$$

$$E_O = \frac{4}{3}E_1$$

$$v_O = \frac{4}{3}v_1$$

So we can relate the current $I_O$ in circuit (b) to the current $I_1$ in circuit (a):

$$i_1 = nAv_1 = nA\left(\frac{3}{4}v_O\right) = \frac{3}{4}nAv_O = \frac{3}{4}i_O$$

(c) Because the mobility u is very high for Cu, only a very small electric field is required to get a large drift speed. Since in the steady state the current through the Cu wires must be the same as the current through the bulb, and since the Cu wires have large cross-sectional area, only a small drift speed is needed in the Cu wires. So E can be very, very small in the connecting wires. Since all the Cu wires have the same cross-sectional area, the same current in all of them $\Rightarrow$ the same drift speed $\Rightarrow$ the same E.

(d) Gray lines are for circuit with two bulbs (part a):

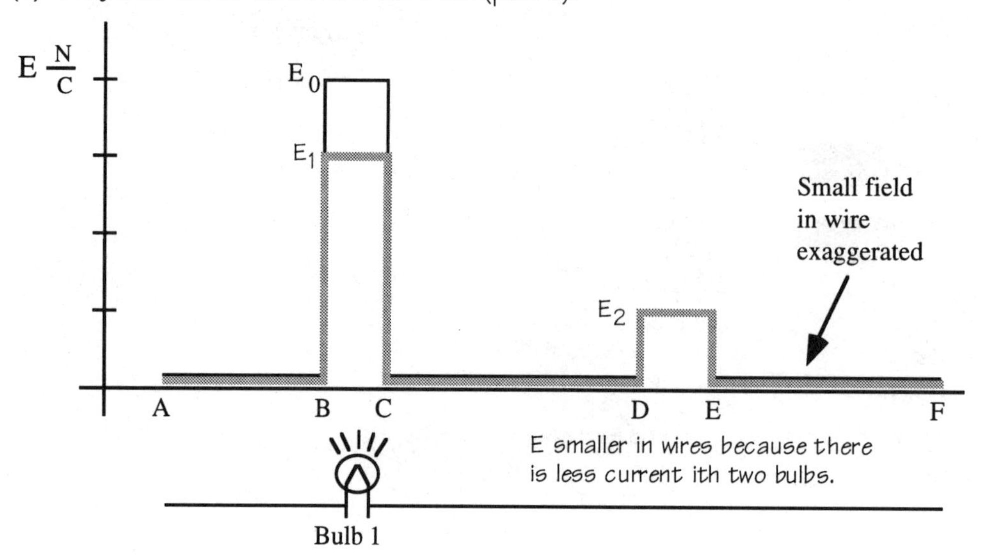


**HW6-5:**

(a)

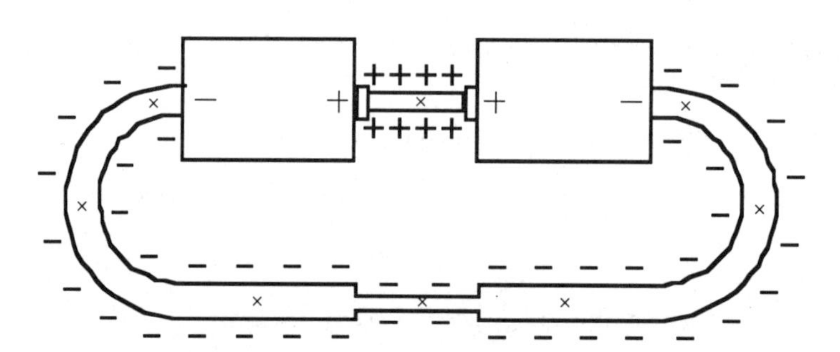

(b) The circuit is in static equilibrium, and E=0 everywhere inside the metal.

(c) The bulb is not glowing: no current.

(d) When we tried this, the bulb didn't light.

## Chapter 7 solutions

**HW7-3:** (a) Compasses point in direction of net magnetic field $\vec{B}$. Contribution of wire at each point is determined by left-hand rule. At point (A) current splits into 2 equal branches, each with half the total current (conservation of charge; definition of steady state, or in this case quasi-steady state, since current is slowly decreasing).

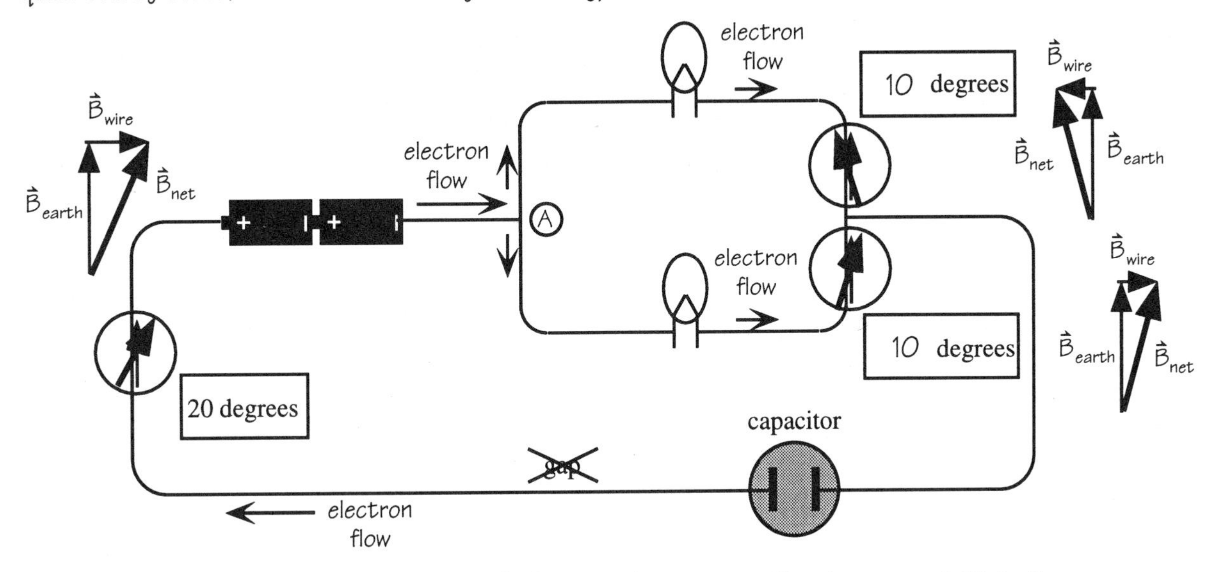


(b) Two bulbs are like having a resistor with 2 times the cross-sectional area, so initially 2 times the current can flow.

| # of bulbs | initial current | charging time |
|---|---|---|
| 1 | $I_1$ | T sec |
| 2 | $2I_1$ | ≈ T/2 sec |

We saw (page 254) that larger initial current led to faster charging, since charge builds up on capacitor faster, leading to greater $E_{fringe}$ in wires, which decreases $E_{net}$ in wires and decreases current flow.

| (c) Experiment: | 2 long bulbs | 1 long bulb |
|---|---|---|
| Time bulbs glowed: | 8 sec; 9 sec | 19 sec; 20 sec |
| Initial compass deflection: | 10 degrees | 4 degrees |

Initial current with 2 bulbs ≈ 2 times initial current with one bulb; $T_{2\ bulbs} \approx T_{1\ bulb}/2$. This is in reasonable agreement with the prediction.

**HW7-4:** 1) Static equilibrium; at location A inside the wire, $\vec{E}_{net} = 0$;

$$\vec{E}_{net} = \vec{E}_{fringe} + \vec{E}_{surface charge} + \vec{E}_{batt} = 0$$

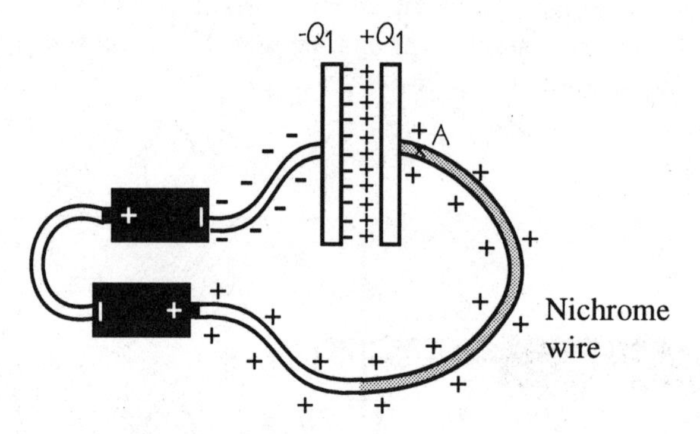


2) Plates closer together; now $|\vec{E}_{fringe}|$ smaller than before, because $E_{fringe}$ proportional to s (distance between plates), so $\vec{E}_{net}$ to left at locations indicated. This leads to more electron flow onto - plate and off of + plate.

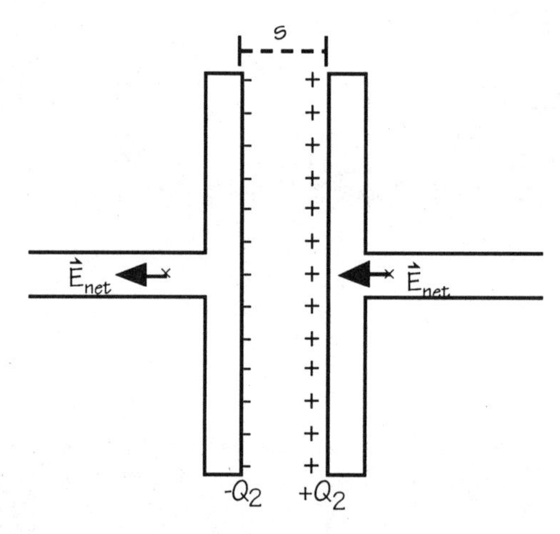


3) New static equilibrium: After some time, enough additional charge builds up on plates so $E_{net} = 0$ again. $E_{fringe}$ increased because Q increased ($E_{fringe}$ proportional to Q).

$$Q_3 > Q_1$$

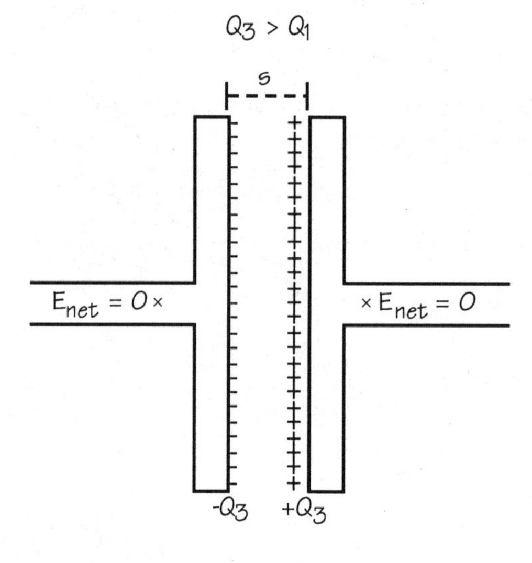

**HW7-5:** 1) Initial transient (first few nanoseconds): $\vec{E}_{cap}$ shown at some key locations, due to charge on metal plates and due to the induced dipoles in the plastic (these two fields are opposite to each other, so $E_{cap}$ is smaller in the wires than it would be without the plastic).

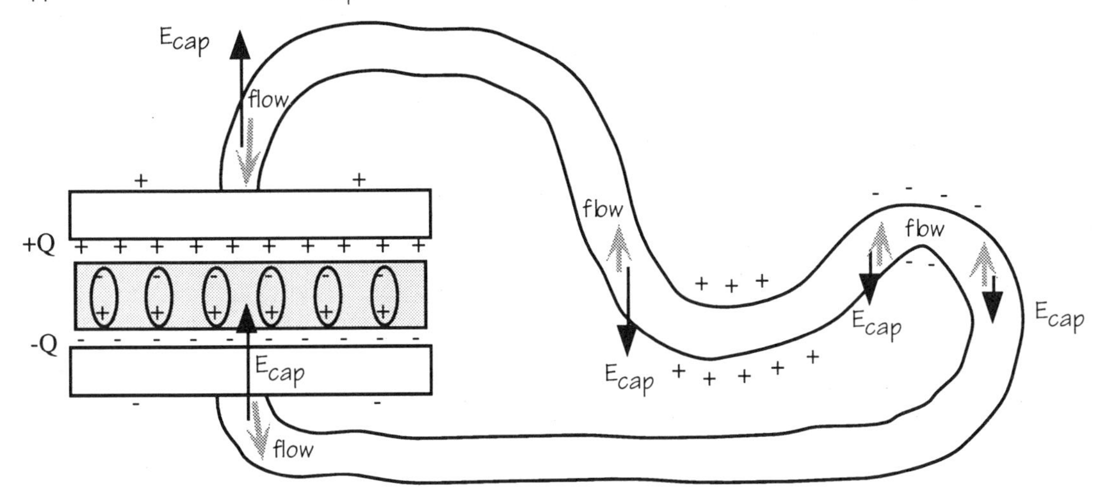


This pattern of $\vec{E}_{cap}$ forces free electrons to flow in the opposite direction to $\vec{E}_{cap}$. On the bends, charge builds up on the surface due to net current entering or leaving a region. The surface charge makes $\vec{E}_{surface}$ in a direction tending toward a steady-state pattern of uniform E throughout the wire, yielding uniform electron current (current conservation in the quasi-steady state, constant cross section for uniform v and E).

2) Quasi-steady state: magnitude E uniform throughout the wire, drives current in the direction to reduce charge on capacitor plates (top plate gets less positive, bottom plate gets less negative). There may be some excess charge on the bends, but the main feature is that there is a gradient of surface charge all along the wire.

As capacitor charge decreases, polarization and surface charge also decrease (proportionally). Hence E decreases $\Rightarrow$ electron current $i = nAuE$ decreases with time (may take 5-10 seconds).

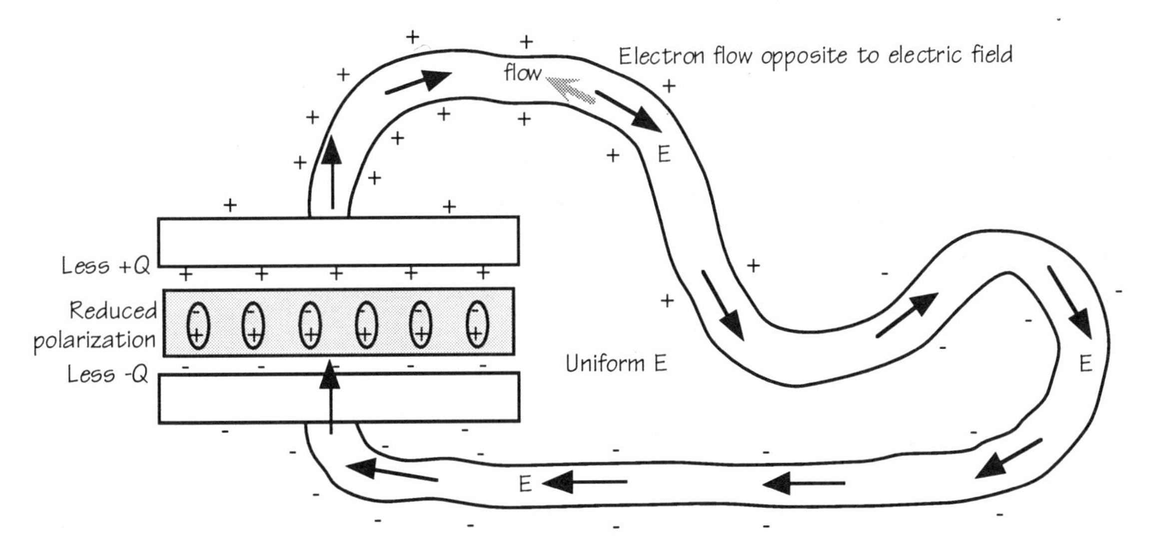


3) Eventually charge on plates reduced to zero, no polarization in plastic, no surface charge on wire: static equilibrium.

## Chapter 8 solutions

**HW8-1:**

(a)

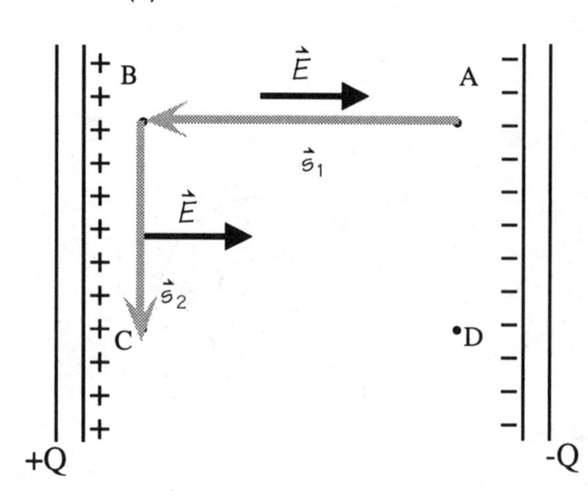


$$\Delta V = V_C - V_A = (V_B - V_A) + (V_C - V_B)$$
$$= -\vec{E}\cdot\vec{s}_1 + -\vec{E}\cdot\vec{s}_2$$
$$= -Es_1\cos 180° + -Es_2\cos 90°$$
$$= +Es_1 + 0$$
$$= +Es_1$$

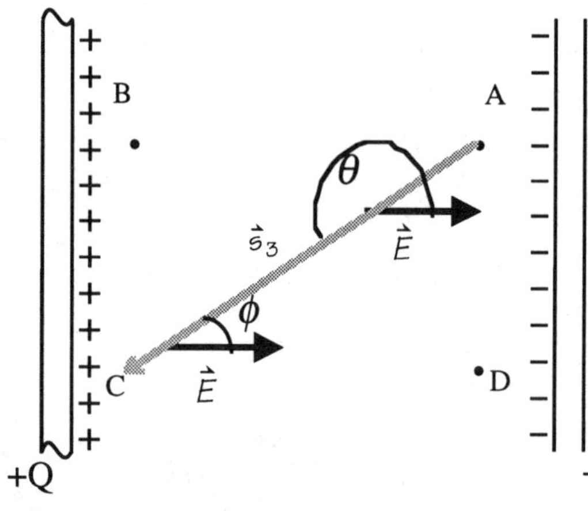


$$\Delta V = V_C - V_A$$
$$= -\vec{E}\cdot\vec{s}_3$$
$$= -Es_3\cos\theta$$
$$= +Es_3\cos\phi \quad \text{(sign change since } \theta = \phi + 180°\text{)}$$
$$s_1 = s_3\cos\phi$$
$$\frac{s_1}{s_3} = \cos\phi$$
$$\Delta V = +Es_3\left(\frac{s_1}{s_3}\right)$$
$$= +Es_1$$

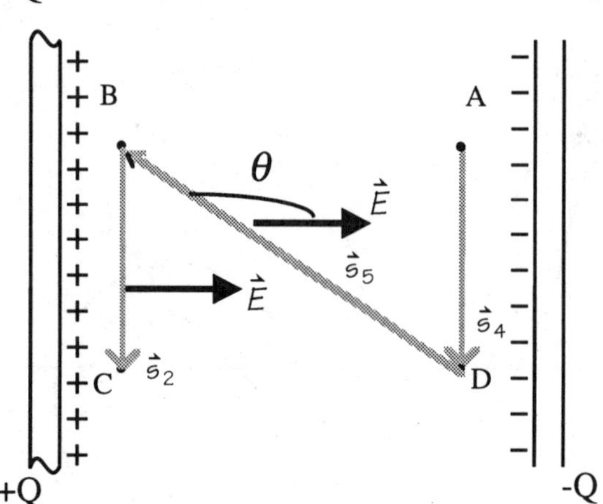


$$\Delta V = V_C - V_A = (V_D - V_A) + (V_B - V_D) + (V_C - V_B)$$
$$= -Es_4\cos 90° + -Es_5\cos\theta + -Es_2\cos 90°$$
$$= 0 + -Es_5\cos\theta + 0$$
$$= -Es_5\left(-\frac{s_1}{s_5}\right) \quad \text{(as on path 2)}$$
$$= +Es_1$$

(b)

$$E = \frac{Q/\pi R^2}{\varepsilon_0} = \frac{(43\times10^{-6}\ C)}{\pi(4\ m)^2\left(9\times10^{-12}\ \frac{C^2}{Nm^2}\right)} = 9.5\times10^4\ \frac{N}{C}$$

$$\Delta V = Es_1 = \left(9.5\times10^4\ \frac{V}{m}\right)\left(1.5\times10^{-3}\ m\right) = 143\ V$$

(c) Consider the path A-B-A.

$$\begin{aligned}\Delta V &= (V_B - V_A) + (V_A - V_B) \\ &= -\vec{E}\cdot\vec{s}_1 + -\vec{E}\cdot(-\vec{s}_1) \\ &= -Es_1 + (+Es_1) \\ &= 0\end{aligned}$$

Consider the path A-D-C-A:

$$\Delta V = (V_D - V_A) + (V_C - V_D) + (V_A - V_D)$$

From part 1:

$$\begin{aligned}&V_D - V_A = 0 \\ &V_C - V_D = +Es_1 \\ &V_A - V_C = -(V_C - V_A) = -Es_1 \\ &\text{So } \Delta V = 0 + +Es_1 + -Es_1 = 0\end{aligned}$$

**HW8-3:**

Fields not drawn to scale on this diagram:

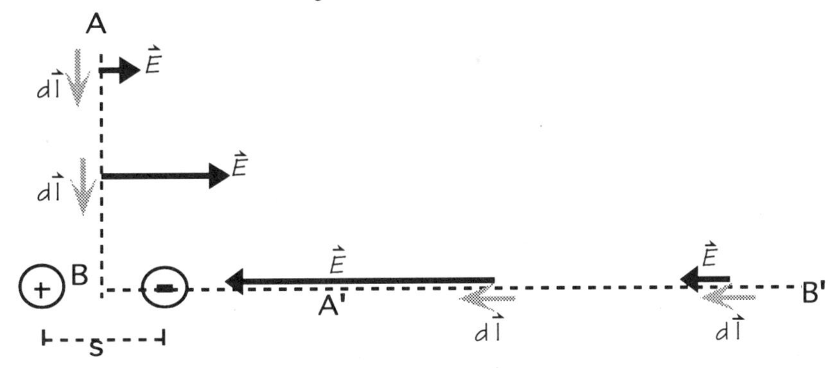


(a) Everywhere on the path from A to B $\vec{E}\perp d\vec{\ell}$, so $\Delta V = -\int_d^0 \vec{E}\cdot d\vec{\ell} = 0$.

$$\begin{aligned}\Delta V = V_{A'} - V_{B'} &= -\int_{B'}^{A'} \vec{E}\cdot d\vec{\ell} = -\int_{B'}^{A'} E\,dl \quad \text{since } \vec{E} // d\vec{\ell} \\ &= -\int_b^a \frac{1}{4\pi\varepsilon_0}\frac{2sq}{x^3}d\ell \quad \text{and since } d\vec{\ell} = -d\vec{x} \\ &= +\int_b^a \frac{1}{4\pi\varepsilon_0}\frac{2sq}{x^3}dx = \frac{2sq}{4\pi\varepsilon_0}\int_b^a \frac{dx}{x^3} \\ &= \frac{2sq}{4\pi\varepsilon_0}\left[-\frac{1}{2}\frac{1}{x^2}\right]_b^a \\ \Delta V &= -\frac{sq}{4\pi\varepsilon_0}\left(\frac{1}{a^2} - \frac{1}{b^2}\right)\end{aligned}$$

$$b > a \text{ so } \frac{1}{b^2} < \frac{1}{a^2} \text{ and therefore } \left(\frac{1}{a^2} - \frac{1}{b^2}\right) > 0$$

so $\Delta V < 0$, as it should be, since we are traveling with the electric field.

It is very easy to make a sign error in this integral; determine the sign on physical grounds instead of trying to get it right in the math (i.e. it's ok to get the wrong sign, as long as you fix it on physical grounds!)

(c)

$$\Delta PE = q\Delta V = (-e)\left(-\frac{5q}{4\pi\varepsilon_0}\left(\frac{1}{a^2}-\frac{1}{b^2}\right)\right)$$

$\Delta PE > 0$ as it should: an external force must do work on the electron to push it in the direction of the electric field.

**HW8-4:**

(a)

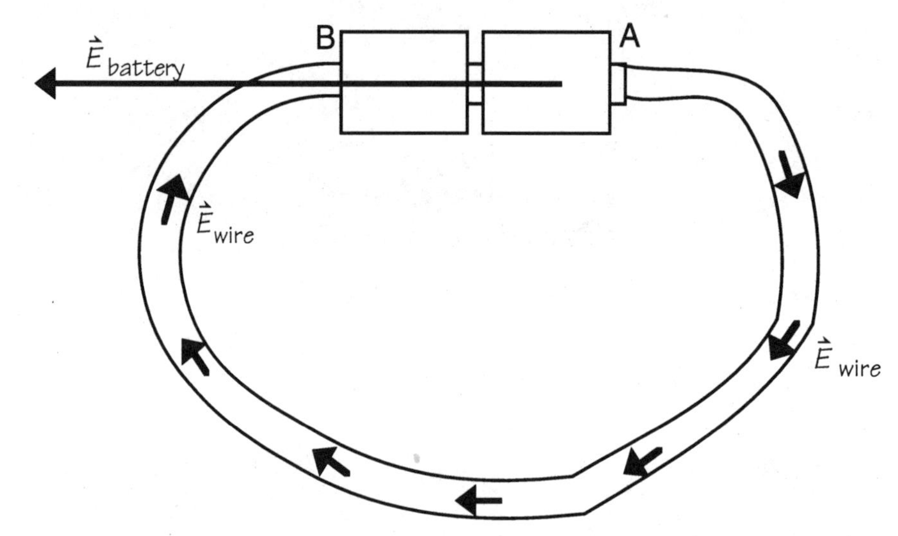


(b) $\Delta V = V_A - V_B = +3.0\ V$ since we are travelling against the electric field (either path).

(As the label on the battery says, $\Delta V$ across one battery is 1.5 V)

(c) Since the electric field in the wire is constant in magnitude and direction (relative to the wire):

$$\Delta V = -EL\cos 0^\circ = -EL$$

$$E = \frac{-\Delta V}{L} = \frac{+3\ V}{0.6\ m} = 5\frac{V}{m}$$

Note: A D-cell is about 6 cm long, so:

$$E_{battery} \approx \frac{3\ \text{volts}}{.12\ m} = 25\ \frac{V}{m}$$

**HW8-5:**

(a)

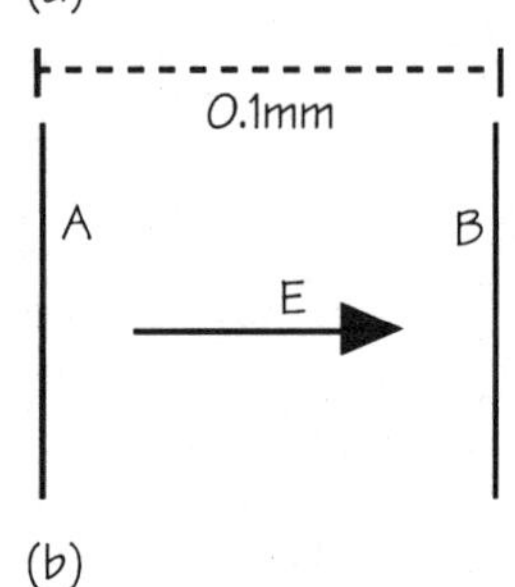


$V_B - V_A < 0$, so $V_B < V_A$

So the electric field must point along the direction from A to B, from higher to lower potential.

This could be due to + charges on the left plate, or - charges on the right plate, or both.

(b)

$$\Delta V = Es\cos 0^\circ = Es$$

$$E = \frac{|\Delta V|}{s} = \frac{10\ V}{10^{-4}\ m} = 10^5\ \frac{V}{m}$$

This electric field is due to the two charged plates. There could be many possible charge configurations; let's assume $|Q_A|=|Q_B|$. Then:

$$E_{capacitor} = \frac{Q/A}{\varepsilon_0}, \text{ or } \varepsilon_0 E_{capacitor} = \frac{Q}{\pi R^2}$$

Only the ratio is fixed; we can pick an arbitrary R and find Q:

For example, let R = 4m (this guarantees that R >> s, so we can use the formula above for E).

$$Q = \varepsilon_0 E \pi R^2 = \left(9\text{x}10^{-12}\ \frac{C^2}{Nm^2}\right)\left(10^5\ \frac{N}{C}\right)\left(\pi(4\ m)^2\right) = 4.5\text{x}10^{-5}\ C$$

**HW8-7:**

Why should there be a <u>maximum</u> potential with respect to infinity for a charged sphere?

The potential of an object relative to infinity depends on the electric field along the path from infinity to the object. We know that there is a maximum electric field: E cannot exceed $3\text{x}10^6$ V/m anywhere along this path without causing breakdown of air. The electric field decreases with distance, so it would be greatest at the surface of the sphere.

Knowing the maximum E at the surface of the sphere, we can find the maximum amount of charge on the sphere

$$E_{at\ surface} = 3\text{x}10^6 \frac{N}{C} = \frac{1}{4\pi\varepsilon_0}\frac{Q_{max}}{R^2}$$

$$Q_{max} = \left(3\text{x}10^6 \frac{N}{C}\right)(4\pi\varepsilon_0)(R^2)$$

and then we can calculate the maximum potential of the sphere relative to infinity

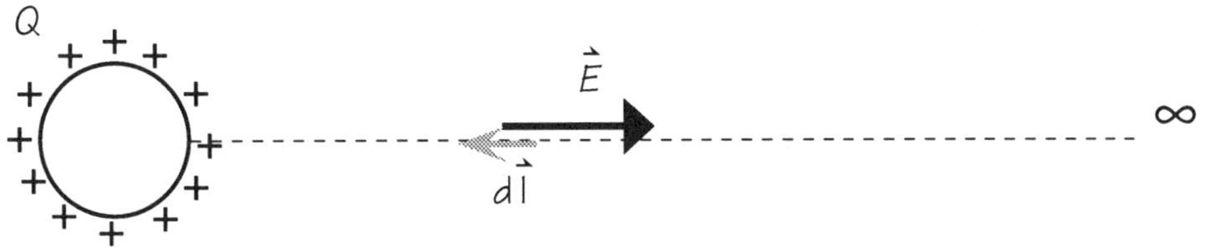

Q    $\vec{E}$    ∞    $d\vec{l}$

$$V_R - V_\infty = -\int_\infty^R \frac{1}{4\pi\varepsilon_0}\frac{Q_{max}}{r^2}dr = \frac{1}{4\pi\varepsilon_0}Q_{max}\left[\frac{1}{R}-\frac{1}{\infty}\right] = \frac{1}{4\pi\varepsilon_0}\frac{Q_{max}}{R}$$

$$= \frac{1}{4\pi\varepsilon_0}\frac{1}{R}\left[\left(3\text{x}10^6 \frac{V}{m}\right)(4\pi\varepsilon_0)(R^2)\right] = \left(3\text{x}10^6 \frac{V}{m}\right)R$$

$$V_{10\ cm} = \left(3\text{x}10^6 \frac{V}{m}\right)(0.1\ m) = 3\text{x}10^5\ V$$

$$V_{1\ mm} = \left(3\text{x}10^6 \frac{V}{m}\right)\left(10^{-3}\ m\right) = 3\text{x}10^3\ V$$

(Actually, for very small spheres it turns out that $E_{critical} > 3\text{x}10^6$ V/m)

**HW8-11:**

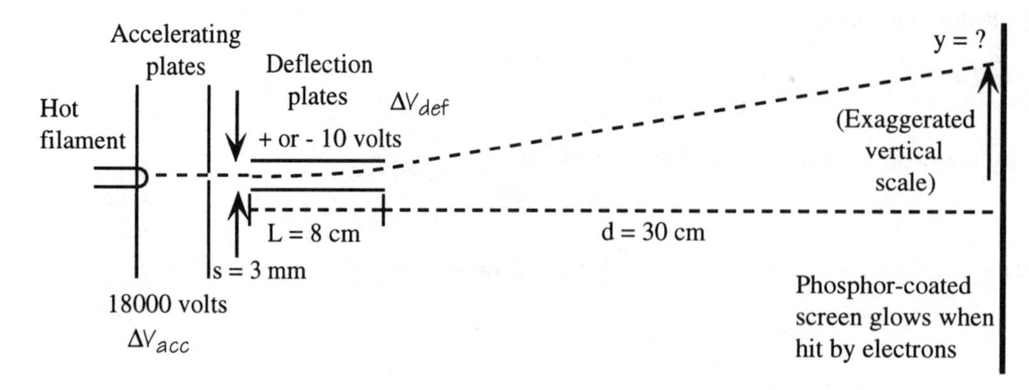


The electron leaves the filament essentially at rest. It is accelerated to the right through a potential difference $\Delta V_{acc}$, and emerges with a velocity $v_x$ in the x-direction. Next it passes through two horizontal deflection plates. During the time it is between these plates, it experiences a force in the +y direction due to the electric field between the plates. It emerges with velocity components $v_x$ (unchanged) and $v_y$. Assuming that the fringe fields of both the accelerating plates and the deflection plates are negligibly small, the electron then travels in a straight line until it hits the screen.

1) x-acceleration:

$$\Delta KE_{electron} = -\Delta PE_{electron}$$

$$\Delta\left(\frac{1}{2}mv_x^2\right) = (e\Delta V_{acc}) \qquad (\Delta PE < 0, \text{ so } -\Delta PE > 0)$$

$$v_x = \sqrt{\frac{2e\Delta V_{acc}}{m}}$$

2) Time to travel through deflection plates:

$$t_{plates} = \frac{L}{v_x}$$

3) E inside deflection plates:

$$\Delta V_{def} = -\vec{E}_{def} \cdot \vec{s} \Rightarrow \left|\vec{E}_{def}\right| = \left|\frac{\Delta V_{def}}{s}\right|$$

4) y-acceleration:

$$a_y = \frac{F}{m} = \frac{eE_{def}}{m}$$

5) v-velocity:

$$v_y = a_y t_{plates}$$

6) Assume that $\Delta y$, the y-distance traveled while inside the deflection plates, is negligibly small compared to y-distance traveled after leaving deflection plates. Now relate $v_x$ and $v_y$:

θ

$y = v_y t$

$d = 30 \text{ cm} = vxt$

$$\tan(\theta) = \frac{v_y}{yx} = \frac{y}{30\text{cm}}$$

Now evaluate this ratio in terms of the other given quantities:

$$\frac{y}{30\text{cm}} = \frac{v_y}{v_x} = \frac{a_y t_{plates}}{v_x} = \frac{\left(\frac{eE_{def}}{m}\right)\left(\frac{L}{v_x}\right)}{v_x} = \frac{\left(\frac{e}{m}\frac{\Delta V_{def}}{s}\right)L}{v_x^{\,2}} = \frac{\left(\frac{e\Delta V_{def}}{m}\right)}{\left(\frac{2e\Delta V_{acc}}{m}\right)}\left(\frac{L}{s}\right) = \frac{1}{2}\frac{\Delta V_{def}}{\Delta V_{acc}}\left(\frac{L}{s}\right) = \frac{1}{2}\frac{(100\text{ V})}{(18000\text{ V})}\left(\frac{L}{s}\right)$$

$$y = (30\text{cm})\frac{1}{2}\frac{(100\text{ V})}{(18000\text{ V})}\left(\frac{8\text{cm}}{0.3\text{cm}}\right) = 2.2\text{cm}$$

It was a good idea to wait until the end to put in numbers, since most of the quantities dropped out, and we avoided lots of unnecessary calculations!

Let's make sure that the electron misses the top deflection plate, and that $\Delta y$, the y-distance traveled inside the deflection plates, is in fact negligible compared to $y = 2.2\text{cm}$, as we assumed.

$$v_x = \sqrt{\frac{2e\Delta V_{acc}}{m}} = \sqrt{\frac{2\left(1.6\times10^{-19}\text{C}\right)\left(1.8\times10^{4}\text{ volts}\right)}{\left(9\times10^{-31}\text{kg}\right)}} = 8\text{x}10^{7}\ \frac{\text{m}}{\text{s}}$$

$$t_{plates} = \frac{L}{v_x} = \frac{8\text{x}10^{-2}\text{m}}{8\text{x}10^{7}\ \frac{\text{m}}{\text{s}}} = 10^{-9}\ \text{s}$$

$$\Delta y = \frac{1}{2}a_y t_{plates}^2 = \frac{1}{2}\frac{e\frac{\Delta V_{def}}{s}}{m}t_{plates}^2 = \frac{\left(1.6\times10^{-19}\text{C}\right)(10\text{volts})}{2\left(9\times10^{-31}\text{kg}\right)\left(3\times10^{-3}\text{m}\right)}\left(10^{-9}\ \text{s}\right)^2 = 3\text{x}10^{-4}\ \text{m}$$

which is less than $\frac{s}{2} = 1.5\text{x}10^{-3}\text{m}$

So the electron does not hit the top plate, and $\Delta y << y$.

**HW8-13:**

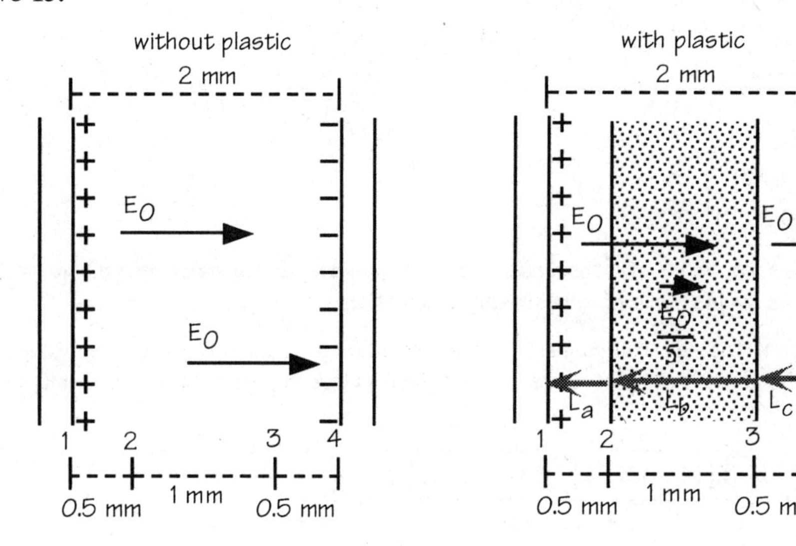


Originally (without plastic), $V_O = 1000$ V, so $E_O = \frac{\Delta V_O}{s} = \frac{1000\ \text{V}}{2\times 10^{-3}\ \text{m}} = 5\times 10^5\ \frac{\text{V}}{\text{m}}$.

When the plastic is added, it polarizes.

*In the air gap beside the plastic,* the net field is now the superposition of the field due to the plates, which has not changed, and the field due to the plastic. In this region, outside the plastic, the field due to the plastic is very small, so we can neglect it. The field in the air gap is therefore unchanged, and $\approx E_O$.

*Inside the plastic,* the polarized molecules of the plastic produce a field to the left. The net field is the superposition of this field, due to the polarized plastic, and the original field due to the plates. The net field inside the plastic is still to the right ($E_{insulator} < E_O$), but is smaller, with magnitude

$$E' = \frac{E_O}{K} = \frac{5\times 10^5\ \frac{\text{V}}{\text{m}}}{5} = 1\times 10^5\ \frac{\text{V}}{\text{m}}$$

($V_1$ - $V_2$) is $\Delta V$ on a path from 2 to 1, so $L_a$ points to the left, and $\Delta V$ should be positive because we are traveling against the electric field:

$$V_1 - V_2 = -\vec{E}_O\cdot\vec{L}_a = -E_O L_a \cos 180^\circ = +E_O L_a = \left(5\times 10^5\ \frac{\text{V}}{\text{m}}\right)\left(5\times 10^{-3}\ \text{m}\right) = +250\ \text{V}$$

Likewise, $L_b$ and $L_c$ point to the left, and ($V_2$-$V_3$) and ($V_3$-$V_4$) should be positive:

$$V_2 - V_3 = -\frac{1}{5}\vec{E}_O\cdot\vec{L}_b = -\frac{1}{5}E_O L_b \cos 180^\circ = +\frac{E_O}{5}L_b = +\left(1\times 10^5\ \frac{\text{V}}{\text{m}}\right)\left(1\times 10^{-3}\text{m}\right) = +100\ \text{V}$$

$$V_3 - V_4 = -\vec{E}_O\cdot\vec{L}_c = +E_O L_c = \left(5\times 10^5\ \frac{\text{V}}{\text{m}}\right)\left(5\times 10^{-3}\ \text{m}\right) = +250\ \text{V}$$

$$V_1 - V_4 = (V_1 - V_2) + (V_2 - V_3) + (V_3 - V_4) = 250\ \text{V} + 100\ \text{V} + 250\ \text{V} = +600\ \text{V}$$

**HW8-16:**

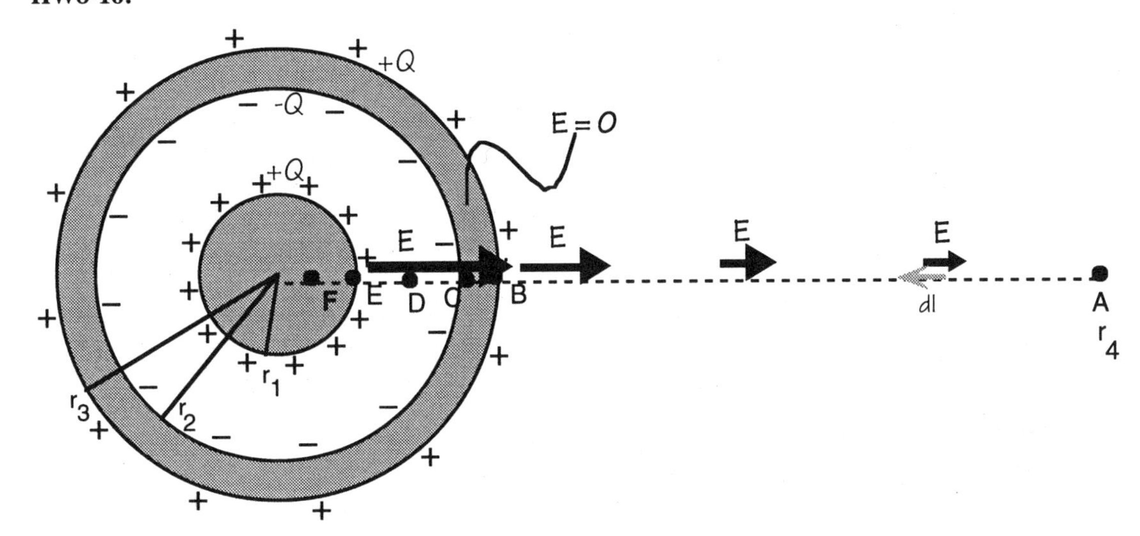


(a) Because all three surfaces have spherically symmetric charge distributions, each surface looks like a point charge located at the origin, as long as we are outside the sphere.

$V_B - V_A > 0$

For r > r3, the net electric field is the superposition of the contributions of all three spheres of charge:: $\vec{E} = \frac{1}{4\pi\varepsilon_0}\frac{+Q}{r^2}\hat{r} + \frac{1}{4\pi\varepsilon_0}\frac{-Q}{r^2}\hat{r} + \frac{1}{4\pi\varepsilon_0}\frac{+Q}{r^2}\hat{r} = \frac{1}{4\pi\varepsilon_0}\frac{+Q}{r^2}\hat{r}$, radially outward, as shown on the diagram. Traveling from A to B we are moving against the electric field, so the potential must increase.

$V_C - V_B = 0$

Inside a metal at static equilibrium, E=0, so $\Delta V = -\vec{E}\cdot\vec{L} = 0$.

$V_D - V_C > 0$

For r1 < r < r2, we are inside two of the spheres of charge, so their contribution to the net electric field is 0. Only the inner sphere gives a nonzero contribution, and $\vec{E} = \frac{1}{4\pi\varepsilon_0}\frac{Q}{r^2}\hat{r}$, radially outward.

Traveling from C to D we are moving against the electric field, so the potential must increase.

$V_F - V_E = 0$

Inside a metal at static equilibrium E=0.

(b)

$$V_F - V_\infty = (V_B - V_\infty) + (V_C - V_B) + (V_E - V_C) + (V_F - V_E)$$

$$= -\int_\infty^B \frac{1}{4\pi\varepsilon_0}\frac{Q}{r^2}dr - \int_B^C 0\,dr - \int_C^E \frac{1}{4\pi\varepsilon_0}\frac{Q}{r^2}dr - \int_E^F 0\,dr$$

$$= +\frac{1}{4\pi\varepsilon_0}Q\left(\frac{1}{r_3} - \frac{1}{\infty}\right) + 0 + \frac{1}{4\pi\varepsilon_0}Q\left(\frac{1}{r_1} - \frac{1}{r_2}\right) + 0$$

$$V_F - V_\infty = \frac{1}{4\pi\varepsilon_0}Q\left(\frac{1}{r_3} + \frac{1}{r_1} - \frac{1}{r_2}\right)$$

$V_F - V_\infty > 0$ as it should be ($r_1 < r_2$ so term in parentheses is positive).

**HW8-17:**

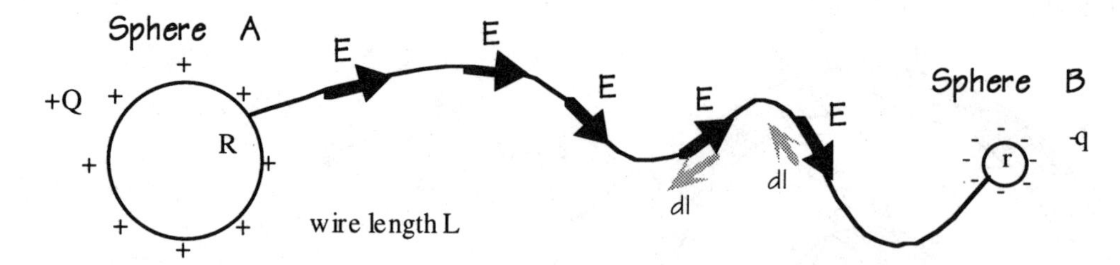


PLAN:

This is a circuit, even though it looks a little odd. If we can calculate an electron current i, we can multiply by $\Delta t$ to find the number of electrons flowing off sphere B in a short time $\Delta t$. We know that $i = nAuE$, and we know n, A, and u, but not E, so we need to find E.

**Finding E in the wire:**

First, since it takes a long time for the circuit to reach equilibrium, it resembles our capacitor circuits, and we can assume that after a few nanoseconds it reaches a "quasi" steady-state in which E is constant throughout the wire, and i is also constant throughout the wire.

The electric field in the wire is the superposition of the fields due to both spheres and to the surface charge on the wire. It is NOT a good approximation to say that E is just the superposition of the fields due to the spheres, because these fields vary a lot with distance. For example, in the middle of the wire, far from both spheres, their contribution must be nearly zero, and the surface charge must be responsible for most of the net field.

Since E is constant along the wire, $\Delta V = V_A - V_B = EL$. So if we knew $V_B - V_A$ we could calculate E.

**Finding $V_B - V_A$:**

Since the two spheres are far apart, and the amount of charge on the wire is very small, the potential of each sphere relative to infinity is approximately the same as if it were an isolated point charge: $V_A - V_\infty = \frac{1}{4\pi\varepsilon_0}\frac{Q}{R}$ and $V_B - V_\infty = \frac{1}{4\pi\varepsilon_0}\frac{-q}{r}$

And $V_A - V_B = (V_A - V_\infty) - (V_B - V_\infty)$, so travelling along the wire from B to A:

$$\Delta V = V_A - V_B = \frac{1}{4\pi\varepsilon_0}\frac{Q}{R} - \frac{1}{4\pi\varepsilon_0}\frac{-q}{r} = \frac{1}{4\pi\varepsilon_0}\left(\frac{Q}{R} + \frac{q}{r}\right), \text{ so}$$

$$E = \frac{\Delta V}{L} = \frac{1}{L}(V_A - V_B) = \frac{1}{L}\left(\frac{1}{4\pi\varepsilon_0}\left(\frac{Q}{R} + \frac{q}{r}\right)\right)$$

**Finding electron current i:**

$$i = nAuE = nAu\frac{1}{L}\frac{1}{4\pi\varepsilon_0}\left(\frac{Q}{R} + \frac{q}{r}\right)\ \frac{\text{electrons}}{\text{second}}$$

**Finding amount of charge that flows in time $\Delta t$:**

In the first few seconds, the charge on each sphere does not change very much, and can be treated as constant. So in time $\Delta t$,

$$\#\ \text{electrons} = i\Delta t = nAuE\Delta t = nAu\frac{\Delta V}{L}\Delta t = \frac{nAu}{L}\frac{1}{4\pi\varepsilon_0}\left(\frac{Q}{R} + \frac{q}{r}\right)\Delta t$$

**HW8-18:**

(a)

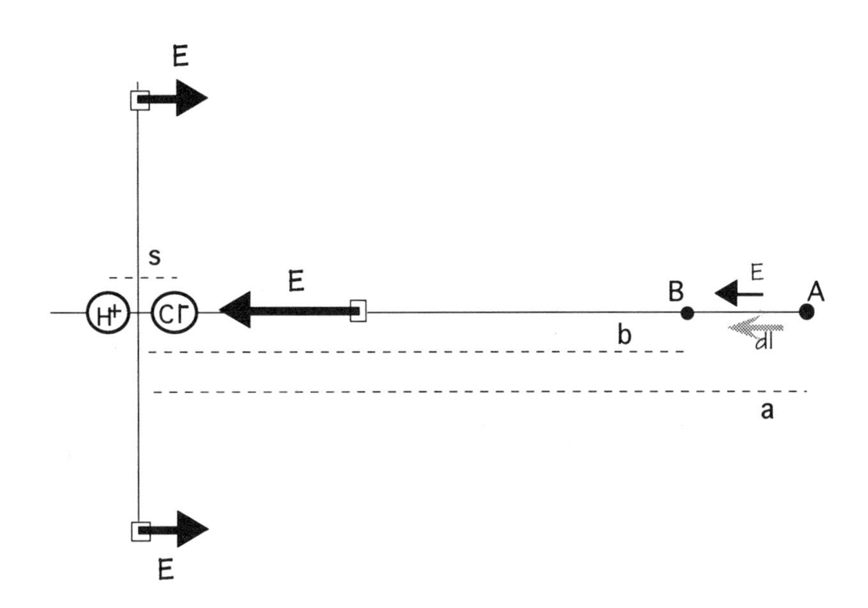


(b)

$$\Delta V = V_B - V_A = -\int_A^B \vec{E}\cdot d\vec{l} = -\int_A^B E dl \cos 0^\circ = -\int_A^B E dl = +\int_A^B E dx \quad (\text{since } dl = -dx)$$

$$= \int_A^B \frac{1}{4\pi\varepsilon_0}\frac{2se}{x^3} = \frac{2se}{4\pi\varepsilon_0}\int_A^B \frac{dx}{x^3} = \frac{2se}{4\pi\varepsilon_0}\left[-\frac{1}{2x^2}\right]_a^b = \frac{se}{4\pi\varepsilon_0}\left(\frac{-1}{b^2} - \frac{-1}{a^2}\right)$$

$$\Delta V = \frac{se}{4\pi\varepsilon_0}\left(\frac{1}{a^2} - \frac{1}{b^2}\right)$$

$\Delta V < 0$ since $a > b$; since we are traveling in the direction of the electric field this is the correct sign.

(c)

Considering the electron plus the HCl molecule as the system, $\Delta KE + \Delta PE = 0$. The change in potential energy is:

$$\Delta PE = q\Delta V = (-e)\frac{2se}{4\pi\varepsilon_0}\left(\frac{1}{a^2} - \frac{1}{b^2}\right) > 0 \text{ since } \Delta V < 0$$

so:

$$KE_B = KE_A + \Delta KE = KE_A - \Delta PE = KE_A - (-e)\frac{2se}{4\pi\varepsilon_0}\left(\frac{1}{a^2} - \frac{1}{b^2}\right)$$

$$KE_B = KE_A + \frac{2se^2}{4\pi\varepsilon_0}\left(\frac{1}{a^2} - \frac{1}{b^2}\right)$$

$$KE_B < KE_A$$

# Chapter 9 solutions

**HW9-2:**

(a)

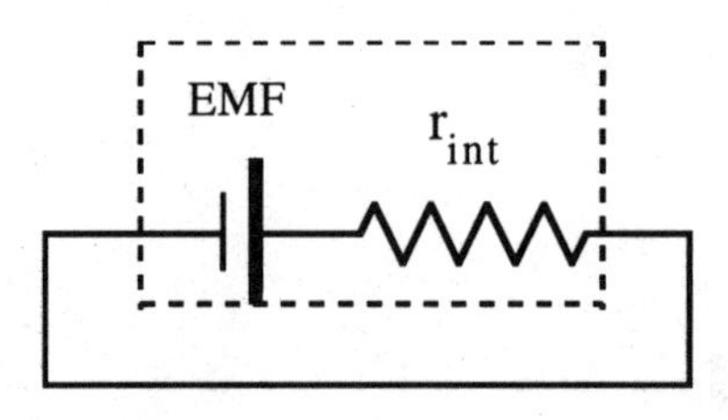


Around the loop:

$$\Delta V_{batt} + \Delta V_{r_{int}} = 0$$

$$EMF - I \cdot r_{int} = 0$$

$$r_{int} = \frac{EMF}{I} = \frac{9V}{18A} = 0.5\Omega$$

(b) Power=$I\Delta V$=(18 A)(9 V)=162 watts generated by the battery.

(c) Power = $I\Delta V$= $I^2R$=(18 A)$^2$ (0.5 $\Omega$)= 162 watts dissipated in the internal resistance or 162 J every second.. This makes sense because in this circuit there are no other resistors, so all the power input by the battery must be dissipated in the internal resistance.

(d)

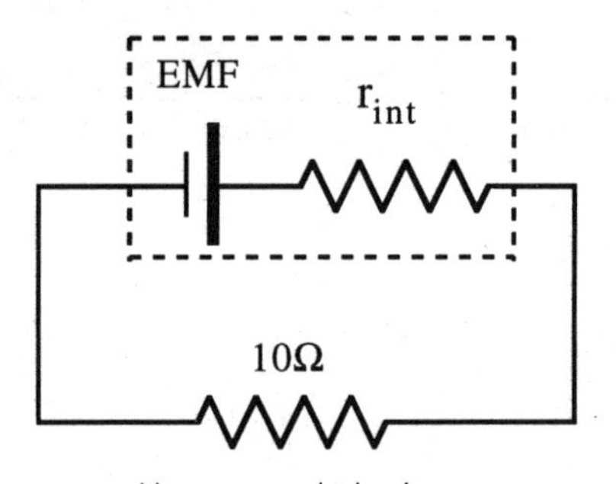


Now around the loop:

$$EMF - I \cdot r_{int} - I \cdot 10\Omega = 0$$

$$I = \frac{EMF}{r_{int} + 10\Omega} = \frac{9\ V}{.5\Omega + 10\Omega} = 0.86\ A$$

(e) Power = $I\Delta V$ = $I(IR)$ = $I^2R$= (0.86 A)$^2$(10$\Omega$)= 7.4 W dissipated in the 10$\Omega$ resistor.

(f)

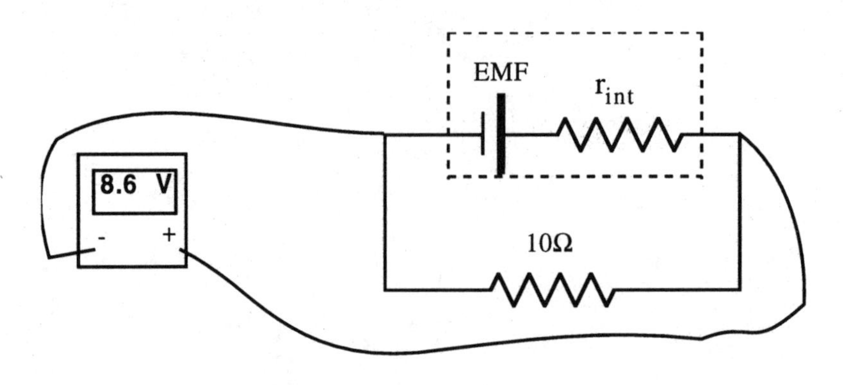


$\Delta V$ = + EMF - I $r_{int}$ = +9 V - (0.86 A)(0.5 $\Omega$) = +8.6 V

**HW9-3:**

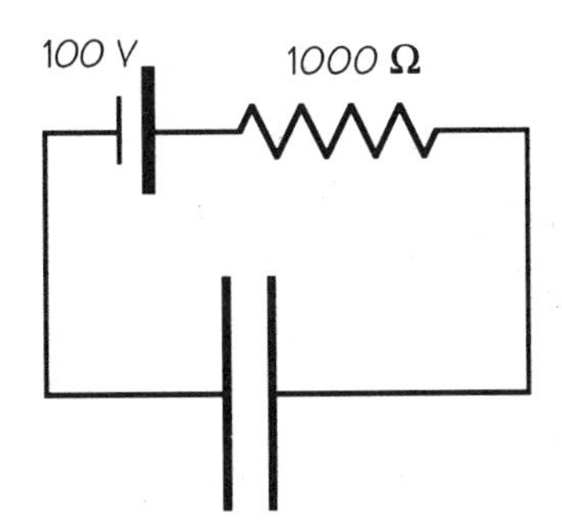


This is an RC circuit, since the deflection plates act as a capacitor. The capacitance of the deflection plates is:

$$C = \frac{Q}{\Delta V} = \frac{Q}{Es} = \frac{Q}{\frac{Q}{A\varepsilon_0}s} = \frac{A\varepsilon_0}{s} = \frac{(.1\text{m})(.02\text{m})\left(9\text{x}10^{-12}\,\text{C}^2/\text{N}\cdot\text{m}^2\right)}{\left(10^{-3}\text{m}\right)} = 1.8\text{x}10^{-11}\ \text{F}$$

The plates charge exponentially:

$$Q = Q_F\left(1 - e^{-\frac{t}{RC}}\right) \quad \text{and} \quad Q = C\Delta V \quad \text{so}$$

$$C\Delta V = C\Delta V_F\left(1 - e^{-\frac{t}{RC}}\right)$$

$$\Delta V = \Delta V_F\left(1 - e^{-\frac{t}{RC}}\right)$$

$$\frac{\Delta V}{\Delta V_F} = \left(1 - e^{-\frac{t}{RC}}\right)$$

$$\left(1 - \frac{\Delta V}{\Delta V_F}\right) = e^{-\frac{t}{RC}}$$

$$\ln\left(1 - \frac{\Delta V}{\Delta V_F}\right) = \ln\left(e^{-\frac{t}{RC}}\right)$$

$$\ln\left(1 - \frac{\Delta V}{\Delta V_F}\right) = -\frac{t}{RC}$$

$$t = -RC\ln\left(1 - \frac{\Delta V}{\Delta V_F}\right) = -\left(10^3\,\Omega\right)\left(1.8\text{x}10^{-11}\ \text{F}\right)\ln\left(1 - \frac{95\text{V}}{100\text{V}}\right)$$

$$t = 5.4\text{x}10^{-8}\ \text{seconds}$$

**HW9-4:** (a)

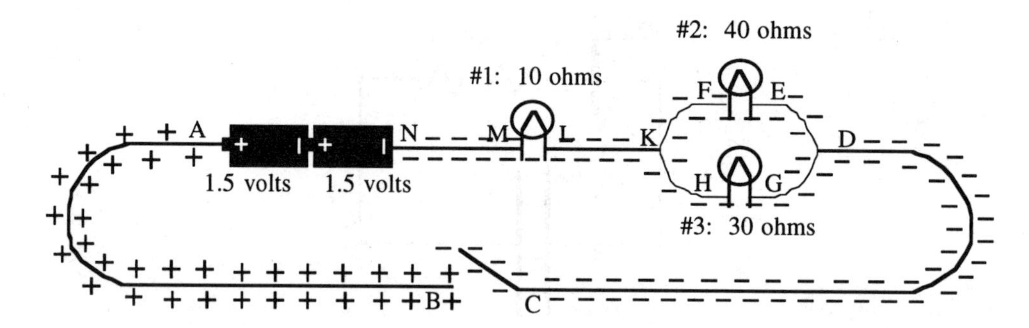


(b) "Walking" from C to B: $V_B - V_C = \Delta V_{wire} + \Delta V_{batteries} + \Delta V_{wire} = 0 + 3V + 0 = 3\ V$

$V_D - V_K = 0$ since the system is at equilibrium and E=0 in the wires.

(c) We need 3 equations since there are 3 unknown currents:

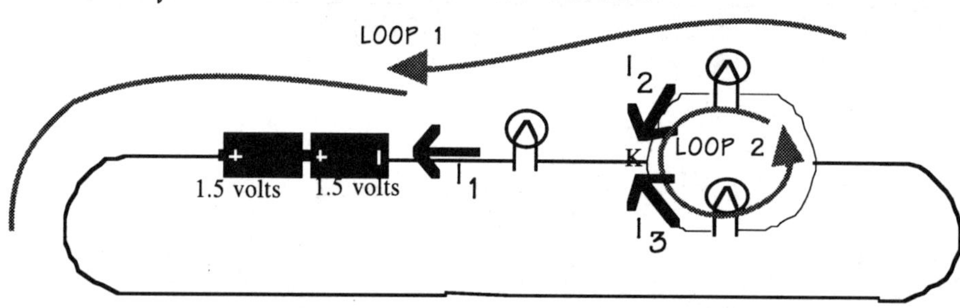


$$\text{Loop 1: } +3\ V - I_2(40\Omega) - I_1(10\Omega) = 0$$

$$\text{Loop 2: } -I_2(40\Omega) + I_3(30\Omega) = 0$$

$$\text{Node K: } I_2 + I_3 = I_1$$

(d) Starting at F and traveling to C: $V_F + I_2(40\Omega) + 0 = V_C$

So $V_C - V_F = I_2(40\Omega)$

(e) Power output of batteries = $I_1(\Delta V) = I_1(3\ V)$

(f) Solving the 3 simultaneous equations:

L1: $+3\ V - I_2(40\Omega) - I_1(10\Omega) = 0$

L2: $-I_2(40\Omega) + I_3(30\Omega) = 0$

$I_2 + I_3 = I_1$

L1: $+3 - 40I_2 - 10(I_2 + I_3) = 0$ (substituting for $I_1$)

L1: $3 - 50I_2 - 10I_3 = 0$

3×L1: $9 - 150I_2 - 30I_3 = 0$

L1+L2: $9 - 190I_2 = 0$

$I_2 = \frac{9}{190}A = .047\ A$

L2: $-(.047)(40) + 30I_3 = 0$

$I_3 = \frac{(40)(.047)}{30} = .063\ A$

$I_1 = I_2 + I_3 = .047 + .063 = .110\ A$

(g) $i = \dfrac{.110\ \frac{\text{coulomb}}{\text{s}}}{1.6\times10^{-19}\ \frac{\text{coulomb}}{\text{electron}}} = 6.88\times10^{17}\ \frac{\text{electron}}{\text{s}}$

(h) $V_C - V_F = I_2(40\Omega) = .047A \times 40\Omega = 1.88\ V$

(i) Power output of batteries = $I_1(\Delta V)$
= $(.110\ A)(3\ V) = .33$ watt

(j) $E_2 = \dfrac{\Delta V_2}{L_2} = \dfrac{I_2R_2}{L_2} = \dfrac{(.047A)(40\Omega)}{(.008m)} = 235\dfrac{V}{m}$

**HW9-9:**

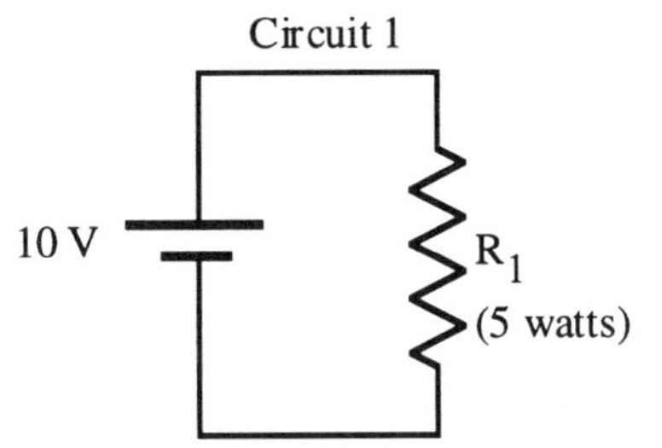


Circuit 2

10 V

$R_2$ (20 watts)

(a)

$$P_1 = 5\text{watts} = I_1\Delta V = I_1 \cdot 10V \qquad P_2 = 20\text{watts} = I_2\Delta V = I_2 \cdot 10V$$

$$I_1 = \frac{5W}{10V} = .5A \qquad I_2 = \frac{20W}{10V} = 2A$$

$$+10V - I_1 \cdot R_1 = 0 \quad \text{(loop eqn)} \qquad +10V - I_2 \cdot R_2 = 0 \quad \text{(loop eqn)}$$

$$R_1 = \frac{10V}{.5A} = 20\Omega \qquad R_2 = \frac{10V}{2A} = 5\Omega$$

(b) $E_1 = \dfrac{\Delta V_1}{L_1} = \dfrac{10V}{3\text{x}10^{-3}m} = 3333\dfrac{V}{m}$

(c)

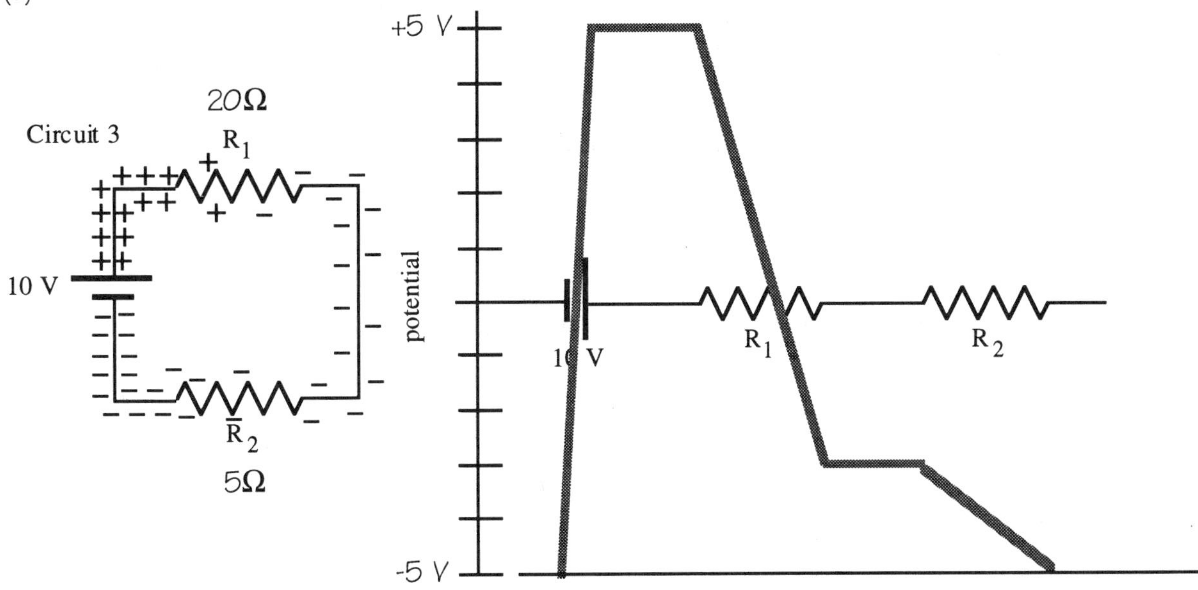


Since $R_1$ contributes 4/5 of the total resistance in the circuit, $\Delta V_1$ will be equal to 4/5 of the EMF. Or,

$$10V - I(20\Omega) - I(5\Omega) = 0$$

$$I = \frac{10V}{25\Omega} = .4A$$

$$\Delta V_1 = (.4A)(20\Omega) = 8V$$

$$\Delta V_2 = (.4A)(5\Omega) = 2V$$

(e) $\#\dfrac{\text{electrons}}{s} = \dfrac{.4\frac{C}{s}}{1.6\text{x}10^{-19}\frac{C}{\text{electron}}} = 2.5\text{x}10^{18}\dfrac{\text{electrons}}{s}$; current is the same everywhere in the series circuit.

(f) $P = I\Delta V = (.4A)(10V) = 4$ watts

## Chapter 10 solutions

**RQ10-1:**

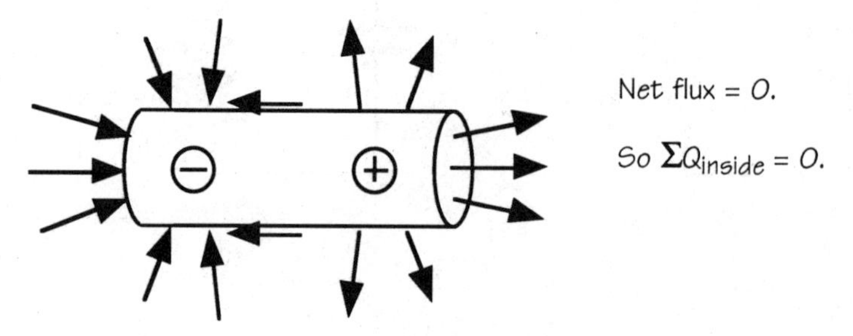

Net flux = 0.

So $\Sigma Q_{inside} = 0$.

**RQ10-2:**

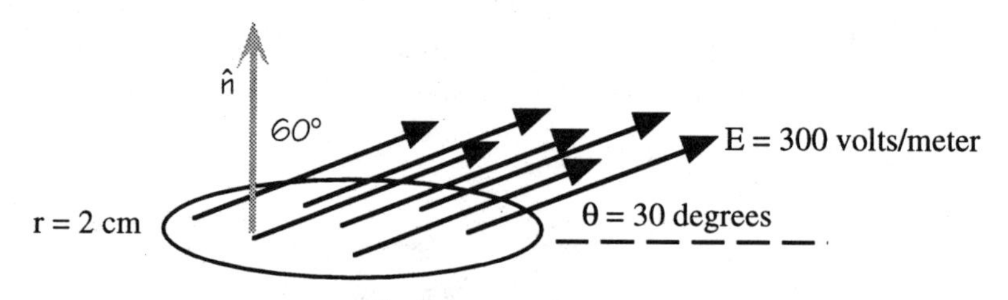


Uniform $\vec{E}$ (magnitude and direction)

$$\text{Flux} = \vec{E} \bullet \hat{n}A = (300V/m)(\cos 60°)\pi(.02m)^2 = 0.19\text{volt}-\text{meter}$$

**RQ10-3:**

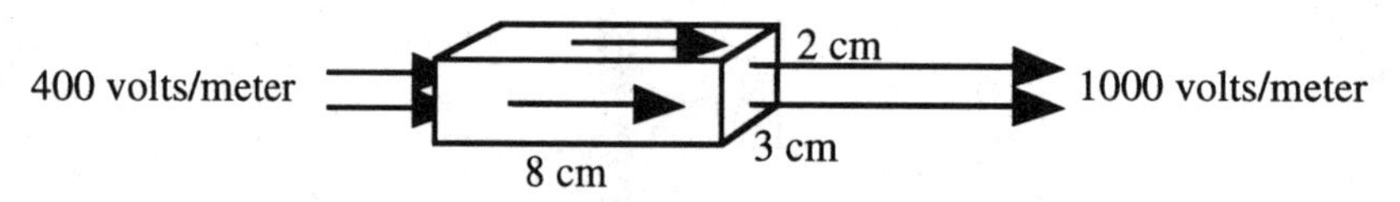


Zero flux on sides because $\vec{E} \bullet \hat{n} = E\cos(90°) = 0$.

Right face: $\vec{E} \bullet \hat{n}A = (1000V/m)(\cos 0°)\left(6 \times 10^{-4}m^2\right) = 0.6\text{volt}-m$

Left face: $\vec{E} \bullet \hat{n}A = (400V/m)(\cos 180°)\left(6 \times 10^{-4}m^2\right) = -0.24\text{volt}-m$

Net flux = [0+0+0+0+0.6+(-0.24)] volt-m = 0.36 volt-m = $\dfrac{\sum Q_{inside}}{\varepsilon_0}$

$$\sum Q_{inside} = (0.36\text{volt}-m)\varepsilon_0 = (0.36\text{volt}-m)\left(9 \times 10^{-12}\frac{C^2}{N\text{-}m^2}\right) = 3.2 \times 10^{-12}C$$

**RQ10-6:**

Car is nearly a closed metal container, so external charges polarize the car in such a way that $\vec{E}_{net} \approx 0$ inside (assuming openings—windows—don't ruin this). No $\vec{E}$, no effect on the body.

But charge on body of car ⇒ $\vec{E}$ not zero outside car, and connecting to ground through body ⇒ current that can kill.

**HW10-1:**

(a) In this part of the problem we reason from what we know about the field to determine what and where the charge must be. To see if there is any charge inside the wire, we draw a (mathematical, imaginary) cylinder of length $\ell$ completely inside the inner wire, and far from the ends of the wire. By Gauss's Law:

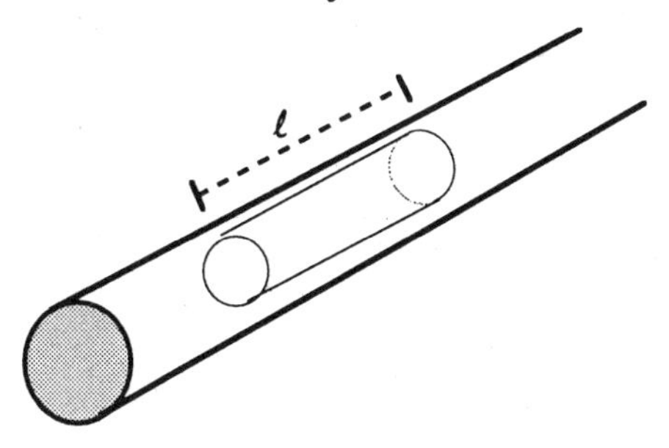


Flux on closed surface = (Charge Inside) / $\varepsilon_0$

$$\int_{\substack{\text{closed}\\\text{cylinder}}} \vec{E}\cdot\hat{n}dA = \frac{\sum q_{\text{inside}}}{\varepsilon_0}$$

At equilibrium E = 0 in metal, so flux on cylinder = 0

$$0 = \frac{\sum q_{\text{inside}}}{\varepsilon_0} \Rightarrow \sum q_{\text{inside}} = 0$$

So there cannot be any charge inside the wire. This means that all the positive charge is on the outer surface of the inner wire.

To see how much of the negative charge is on the inner surface of the outer cylinder, and how much is on the outer surface, we draw a (mathematical) cylindrical surface inside the metal of the outer cylinder, again far from the ends of the wire.

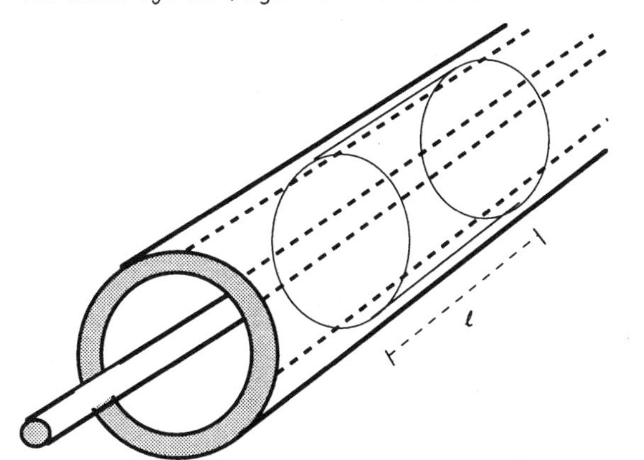


$$\int_{\substack{\text{closed}\\\text{cylinder}}} \vec{E}\cdot\hat{n}dA = \frac{\sum q_{\text{inside}}}{\varepsilon_0}$$

$$\int_{\text{end caps}} \vec{E}\cdot\hat{n}dA + \int_{\text{curved side}} \vec{E}\cdot\hat{n}dA = \frac{\sum q_{\text{inside}}}{\varepsilon_0}$$

The curved (side) part of the cylinder is entirely within the metal, and as before E=0 inside the metal in equilibrium, so flux on the curved part =0. Only part of the end cap is inside the metal; the other part is in the air gap, where there is a nonzero electric field.

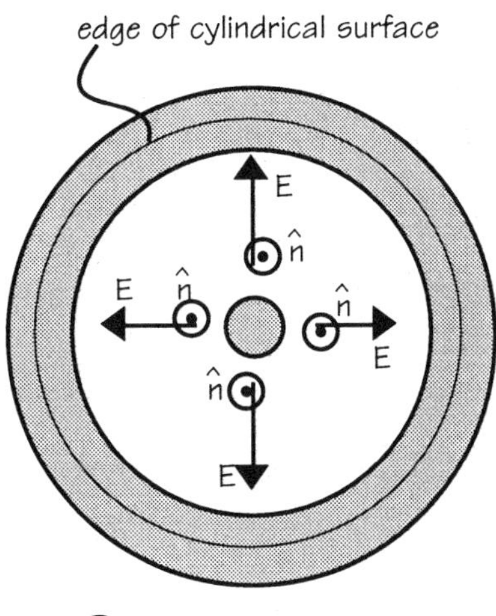


Looking end-on at the cylinder, we see that $\hat{n}$ is out of the page, while $\vec{E}$ points out from the inner wire (by symmetry) and is in the plane of the page, parallel to the surface of the end cap. $\vec{E}$ is therefore perpendicular to $\hat{n}$, and $\vec{E}\bullet\hat{n} = 0$. So the flux on the end caps of the mathematical cylinder is zero, and therefore the net flux on the cylinder is zero.

By Gauss's Law, the net charge enclosed by this cylindrical surface must therefore be 0.

We know that the cylinder encloses a charge $(+Q/L)\ell$ on the inner wire. Thus:

$$0 = (+Q/L)\ell + (\text{density on inner surface})\ell$$

and therefore

$$(\text{density on inner surface}) = (-Q/L).$$

All the negative charge is on the inner surface.

(b) In this part of the problem we reason from what we now know about the charge distribution, plus what we know about the direction and symmetry of $\vec{E}$, to get an algebraic expression for the magnitude of $\vec{E}$ inside the air gap.

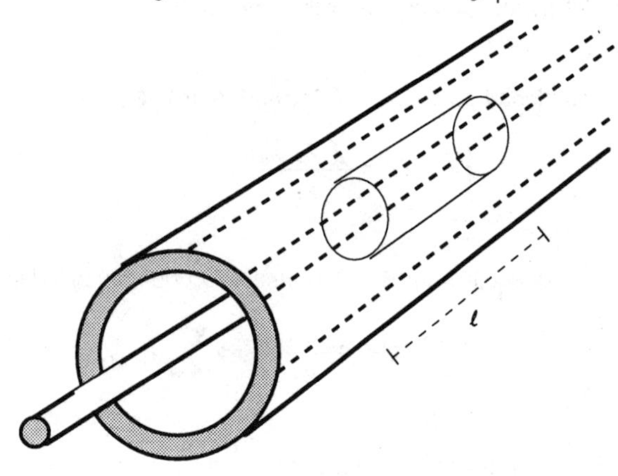


We place our mathematical surface so that at least one side (the curved part) is located in the region where we want to know the electric field—in the air gap. Again:

$$\int_{\text{closed cylinder}} \vec{E}\cdot\hat{n}dA = \frac{\sum q_{inside}}{\varepsilon_0}$$

$$\int_{\text{end caps}} \vec{E}\cdot\hat{n}dA + \int_{\text{curved side}} \vec{E}\cdot\hat{n}dA = \frac{\sum q_{inside}}{\varepsilon_0}$$

By symmetry, the electric field in the air gap must point outward from the wire.

End Caps:

On the end caps, therefore, as before, $\vec{E}\bullet\hat{n} = 0$ and the flux $=0$.

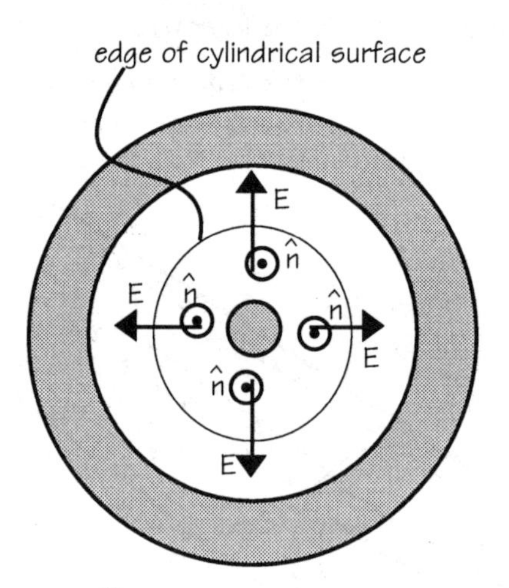


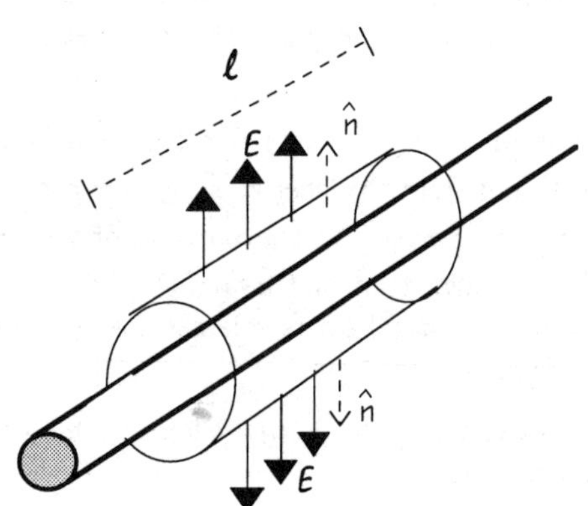


Curved Surface:

E is nonzero, and points outward, parallel to $\hat{n}$, so $\vec{E}\bullet\hat{n} = E$.

On the curved surface, at a constant distance from the wire, the magnitude E is constant, by symmetry.

We can partially evaluate the flux:

$$\int_{\text{curved surface}} \vec{E}\cdot\hat{n}dA = \int_{\text{curved surface}} E\cos 0^\circ\, dA = E\int_{\text{curved surface}} dA = E(2\pi r\ell)$$

(where $2\pi r\ell$ is the area of the side of the cylinder).

The charge inside the cylinder is equal to (charge per unit length)x(length of cylinder):

$$\underset{\text{end caps}}{\int} \vec{E}\cdot\hat{n}dA + \underset{\text{curved side}}{\int} \vec{E}\cdot\hat{n}dA = \frac{\sum q_{\text{inside}}}{\varepsilon_0}$$

$$0 + 2\pi r\ell E = \frac{(+Q/L)\ell}{\varepsilon_0}$$

$$E = \frac{(+Q/L)\ell}{2\pi r\ell\varepsilon_0} = \frac{(+Q/L)}{2\pi\varepsilon_0 r}$$

This is the magnitude of E in the air gap.

(c) Finally we find the magnitude of the electric field outside the whole assembly:

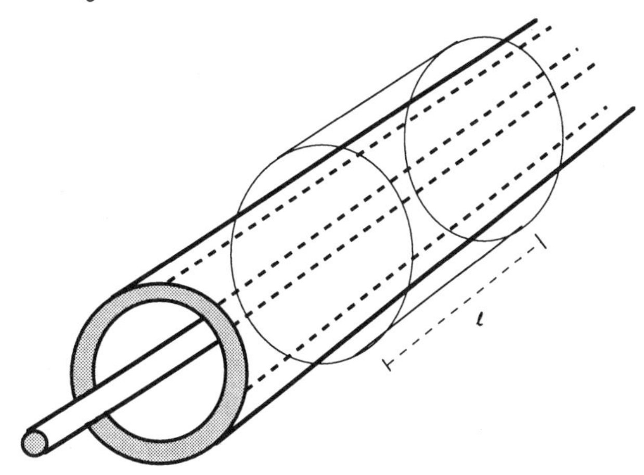


Our closed surface now extends around both wires.

$$\underset{\text{closed surface}}{\int} \vec{E}\cdot\hat{n}dA = \frac{\sum q_{\text{inside}}}{\varepsilon_0}$$

$$\underset{\text{end caps}}{\int} \vec{E}\cdot\hat{n}dA + \underset{\text{curved surface}}{\int} \vec{E}\cdot\hat{n}dA = \frac{(+Q/L)\ell + (-Q/L)\ell}{\varepsilon_0}$$

$$0 + E(2\pi r\ell) = \frac{0}{\varepsilon_0}$$

$\therefore E = 0$ on curved surface

As above, on end caps $\vec{E}\bullet\hat{n} = 0$.

On curved surface E must point outward, so $\vec{E}\bullet\hat{n} = E$, so flux $= E(2\pi r\ell)$

# Chapter 11 solutions

**RQ11-1:**

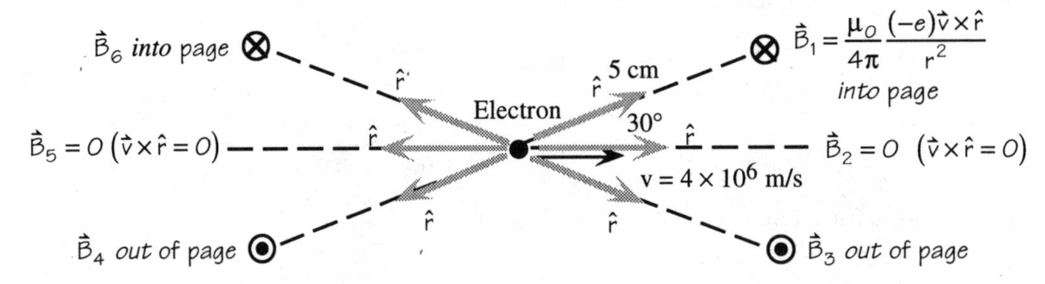


At $P_1$, $P_3$, $P_4$, and $P_6$, $|\vec{v}\times\hat{r}| = v\sin(30°)$ (or $v\sin(150°) = v\sin(30°)$).

$$B_1 = B_3 = B_4 = B_6 = \frac{\mu_0}{4\pi}\frac{ev\sin(30°)}{r^2} = \left(10^{-7}\frac{T-m^2}{C-m/s}\right)\frac{(1.6\times10^{-19}C)(4\times10^6 m/s)(0.5)}{(.05m)^2}$$

$$= 1.3\times10^{-17}\text{ tesla}$$

**RQ11-3:**

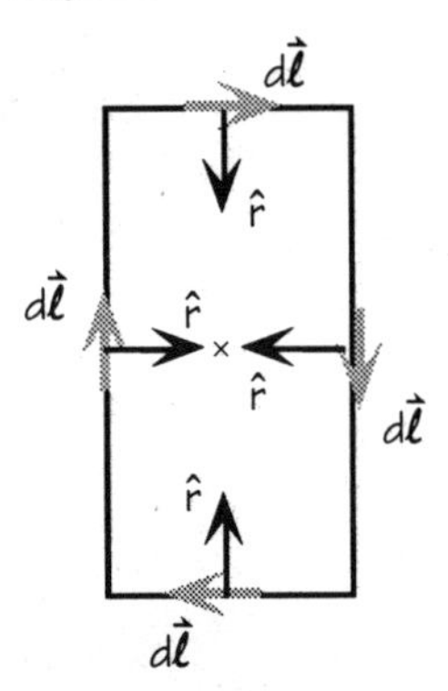


At center, all the $d\vec{l}\times\hat{r}$'s point *into* the page.

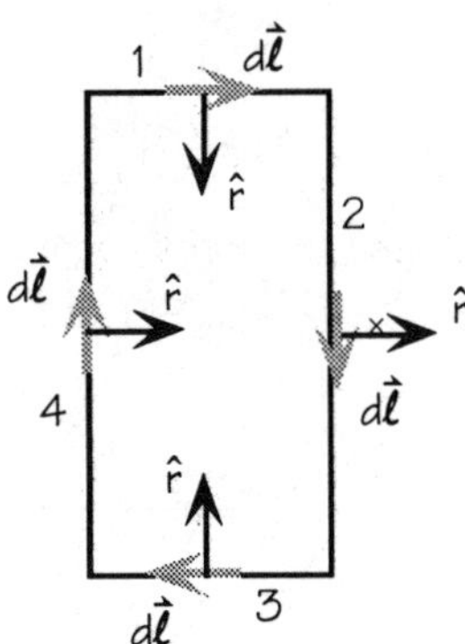


Just outside, all the $d\vec{l}\times\hat{r}$'s for wires 1, 3, and 4 point *into* the page.

But $d\vec{l}\times\hat{r}$ for all segments of wire 2 points *out* of the page, and $B_2$ is the biggest contribution if we are near wire 2. Hence $\vec{B}_{net}$ points *out* of the page.

**RQ11-9:** At a distance from the magnet, the net field is the superposition of the magnetic field due to each iron atom, each of which is approximately the same (long) distance away. Since the 1-kg magnet has 10 times as many atoms as the 100-gram magnet, it will make a magnetic field 10 times as big.

**HW11-1:**

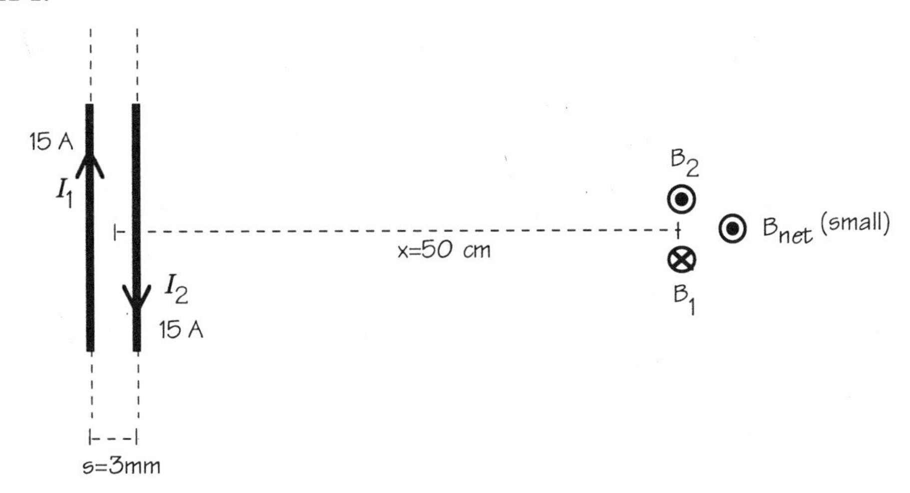


(a) At any instant the wires make magnetic fields in opposite directions; since $I_2$ is closer this results in a small net field.

Assuming the wires are very long:

$$B_{net} \approx \frac{\mu_0}{4\pi}\frac{2I}{\left(x-\frac{s}{2}\right)} - \frac{\mu_0}{4\pi}\frac{2I}{\left(x+\frac{s}{2}\right)} = \frac{\mu_0}{4\pi}2I\left[\frac{1}{\left(x-\frac{s}{2}\right)} - \frac{1}{\left(x+\frac{s}{2}\right)}\right] = \frac{\mu_0}{4\pi}2I\left[\frac{s}{x^2-\left(\frac{s}{2}\right)^2}\right] \approx \frac{\mu_0}{4\pi}\frac{2Is}{x^2}$$

$$\approx \left(10^{-7}\right)\frac{2(15\ \text{A})(0.003\ \text{m})}{(.5\ \text{m})^2}$$

$$B_{net} \approx 3.6\text{x}10^{-8}\ \text{T}$$

(b) If the wires are twisted then their fields should very nearly cancel, since regions where wire 1 is closer are balanced by regions where wire 2 is closer.

**HW11-2:**

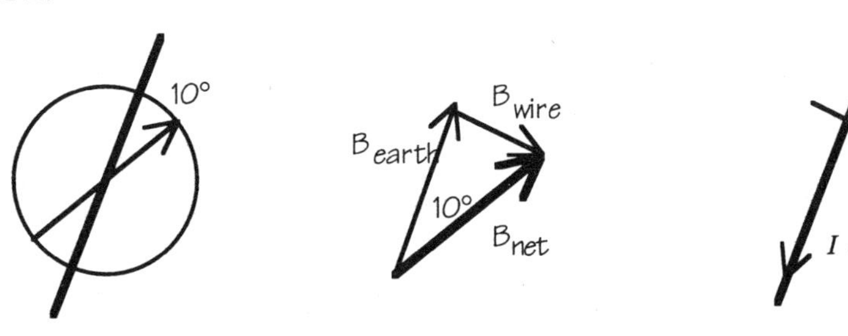


(a) The magnetic field under the wire points to the right, so using the right hand rule (thumb in direction of conventional current, fingers curl around in direction of B) one's thumb must point down, so conventional current flows down (toward the bottom of the page).

(b)

$$B_{wire} = B_{earth}\tan 10^\circ = \left(2\text{x}10^{-5}\ \text{T}\right)\tan 10^\circ = 3.5\text{x}10^{-6}\ \text{T}$$

$$B_{wire} = \frac{\mu_0}{4\pi}\frac{2I}{x} \Rightarrow I = \frac{B_{wire}x}{2\frac{\mu_0}{4\pi}} = \frac{\left(3.5\text{x}10^{-6}\ \text{T}\right)(.005\,\text{m})}{2\left(10^{-7}\right)} = 8.8\text{x}10^{-2}\ \text{A}$$

**HW11-4:**

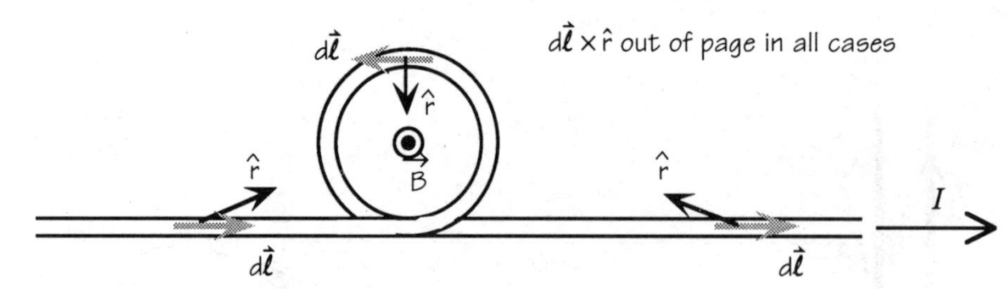


The magnetic field at the center of the ring is the sum of the magnetic field of the straight wire and the magnetic field of the loop. Both the wire and the loop make B out of the page, so the net field points out of the page.

$$\vec{B}_{loop} = \int_{loop} \frac{\mu_o}{4\pi}\frac{I\,d\vec{\ell}\times\hat{r}}{r^2} \qquad \text{and } d\vec{\ell}\perp\hat{r} \text{ everywhere on loop, so}$$

$$B_{loop} = \frac{\mu_o}{4\pi}\frac{I}{r^2}\int_{loop} d\ell \sin 90^\circ = \frac{\mu_o}{4\pi}\frac{I}{r^2}\int_{loop} d\ell = \frac{\mu_o}{4\pi}\frac{I}{r^2}(2\pi r) = \frac{\mu_o}{4\pi}\frac{2\pi I}{r} \quad \text{out of the page}$$

$$B_{long\,wire} = \frac{\mu_o}{4\pi}\frac{2I}{r} \quad \text{out of the page}$$

$$B_{net} = \frac{\mu_o}{4\pi}\frac{2I}{r}(1+\pi) \quad \text{out of the page}$$

**HW11-5:**

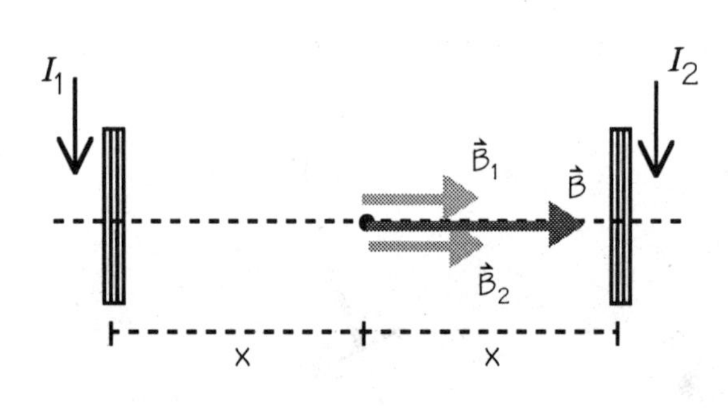


(a) Both coils make a magnetic field in the same direction at the point of interest. Since the point is halfway between them, the magnitudes of the contributions are also the same:

$$B_1 = B_2 = (10)\frac{\mu_o}{4\pi}\frac{2\pi R^2 I}{\left(x^2+R^2\right)^{\frac{3}{2}}}$$

$$B_{net} = B_1 + B_2 = (10)\frac{\mu_o}{4\pi}\frac{4\pi R^2 I}{\left(x^2+R^2\right)^{\frac{3}{2}}}$$

$$= (10)\left(10^{-7}\right)\frac{4\pi(.03\text{m})^2(2\text{A})}{\left((.1\text{m})^2+(.03\text{m})^2\right)^{\frac{3}{2}}}$$

$$B_{net} = 2.0\text{x}10^{-5}\text{ T to the right}$$

(b) Using the $\frac{1}{x^3}$ approximation:

$$B_{net} = (10)2\left(\frac{\mu_o}{4\pi}\frac{2\pi R^2 I}{x^3}\right) = 2(10)\left(10^{-7}\right)\left(\frac{2\pi(.03\text{m})^2(2\text{A})}{(.1\text{m})^3}\right) = 2.3\text{x}10^{-5}\text{ T}$$

$$\frac{2.3}{2.0} = 1.15 \quad \text{so there is a 15\% error}$$

(c) If the current in the right loop is reversed, the magnetic field $B_2$ now points to the left. Since it is still equal in magnitude to $B_1$, the net field is zero at the midpoint.

**HW11-6:**

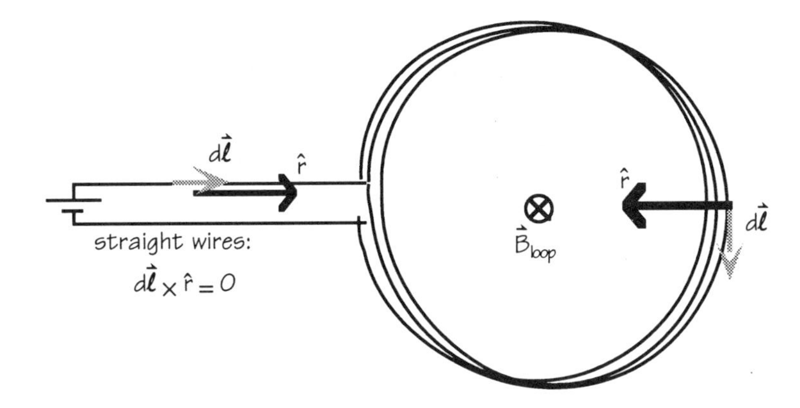


First we need to find the magnetic field due to the current in the circuit, at the position of the compass (center of the loops).

The straight segments of wire make no contribution, since for them $d\vec{\ell} \parallel \hat{r}$.

The magnetic field due to a representative segment of the loop is into the page, at the center of the loop. All segments of the loop contribute $d\vec{B}$ in the same direction. The magnitude of B due to all 3 loops is:

$$B = 3\frac{\mu_0}{4\pi}\frac{2\pi I}{a} = 3\left(10^{-7}\ \frac{T\cdot m}{A}\right)\frac{\left(2\pi\left(\frac{1.5\ V}{6\ \Omega}\right)\right)}{(.15\ m)} = 3\text{x}10^{-6}\ T;$$ B is into the page as shown above.

Second, we need to compute the net magnetic field at the location of the compass. Looking down from above:

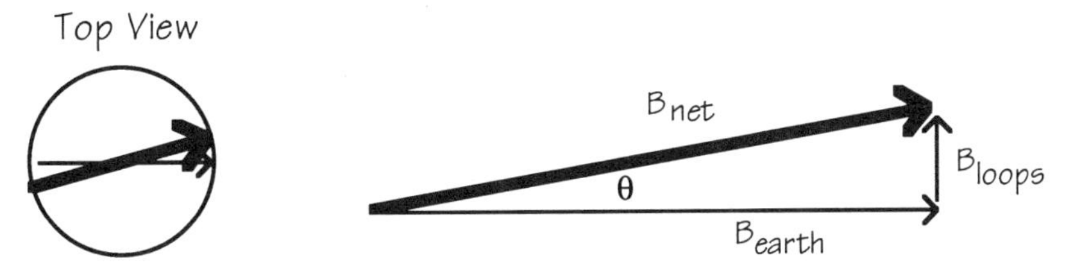


The compass deflects inward. $\theta = \arctan\left(\frac{B_{loops}}{B_{earth}}\right) = \arctan\left(\frac{3\text{x}10^{-6}\ T}{2\text{x}10^{-5}\ T}\right) = 8.5°$

**HW11-7:**

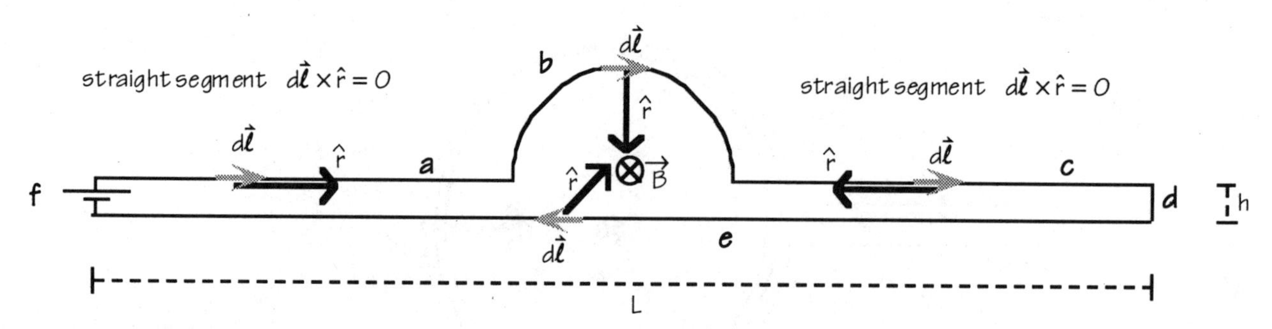


Since "failing to account for the magnetic field contribution of every segment of a circuit is a sin against the superposition principle," we will consider the segments labeled a - f:

Straight segments a and c: Contribute 0, since $d\vec{\ell} \parallel \hat{r}$ or $-d\vec{\ell} \parallel \hat{r}$.

End segments d and f: we can ignore these wires, because they are very short and far away, and their contribution to B falls off like $\frac{1}{d^2}$.

Semicircular segment b:

Every current element contributes dB into the page, as shown above.

$$\left|\vec{B}_{\text{semi circle}}\right| = \left|\int_{\text{semi circle}} d\vec{B}\right| = \left|\int_{\text{semi circle}} \frac{\mu_0}{4\pi}\frac{I d\vec{\ell} \times \hat{r}}{r^2}\right| = \frac{\mu_0}{4\pi}\frac{I}{r^2}\int_{\text{semi circle}} d\ell \sin 90^\circ = \frac{\mu_0}{4\pi}\frac{I}{r^2}(\pi r) = \frac{\mu_0}{4\pi}\frac{\pi I}{r}$$

Straight segment e:

B due to the lower wire is also into the page at the point of interest.

Assuming that the length L of this wire is much greater than h, the distance to the point of interest, we can use the very long wire approximation:

$$B_{\text{long wire}} = \frac{\mu_0}{4\pi}\frac{2I}{h}$$

So the net magnetic field is into the page, with magnitude:

$$B_{\text{net}} = \frac{\mu_0}{4\pi}\frac{\pi I}{r} + \frac{\mu_0}{4\pi}\frac{2I}{h}$$

## Chapter 12 solutions

**HW12-1:** (a)

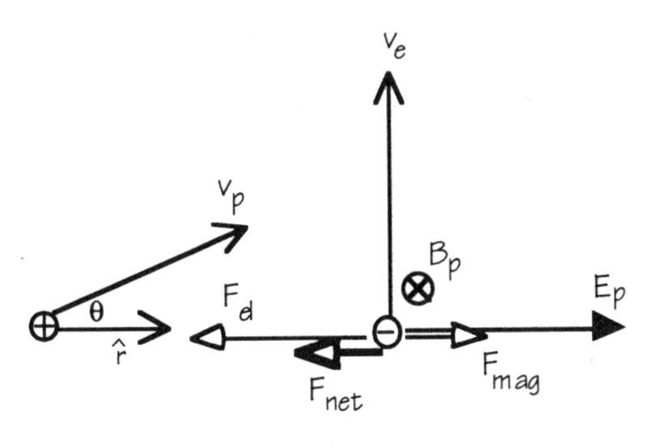


Force on electron:

Proton makes B into page at location of electron.

$$B_p = \left| \frac{\mu_0}{4\pi} \frac{e\vec{v}_p \times \hat{r}}{d^2} \right| = \frac{\mu_0}{4\pi} \frac{e v_p \sin\theta}{d^2} \quad \text{into page}$$

Magnetic force on electron by proton is to right.

$$F_{mag} = \left| -e\vec{v}_e \times \vec{B}_p \right| = e v_e B_p \sin 90° = \frac{\mu_0}{4\pi} \frac{e^2 v_p v_e \sin\theta}{d^2}$$

Electric force on electron by proton:

$$F_{el} = \left| -e\vec{E}_p \right| = \frac{1}{4\pi\varepsilon_0} \frac{e^2}{d^2} \quad \text{to left}$$

Net Force:

$$\vec{F}_{net} = \vec{F}_{el} + \vec{F}_{mag}$$

$$F_x = -\frac{1}{4\pi\varepsilon_0} \frac{e^2}{d^2} + \frac{\mu_0}{4\pi} \frac{e^2 v_e v_p \sin\theta}{d^2}$$

$$= -\frac{1}{4\pi\varepsilon_0} \frac{e^2}{d^2} \left[ 1 - \frac{v_e v_p \sin\theta}{c^2} \right] \quad (\text{since } \mu_0\varepsilon_0 = \frac{1}{c^2})$$

$$F_y = 0$$

(b)

Force on proton:

Electron makes B into page at location of proton.

$$B_e = \left| \frac{\mu_0}{4\pi} \frac{-e\vec{v}_e \times \hat{r}}{d^2} \right| = \frac{\mu_0}{4\pi} \frac{e v_e}{d^2} \quad \text{into page}$$

Magnetic force on proton by electron is to northwest:

$$F_{mag} = \left| e\vec{v}_p \times \vec{B}_e \right| = e v_p B_e \sin 90° = \frac{\mu_0}{4\pi} \frac{e^2 v_p v_e}{d^2}$$

Electric force on proton by electron:

$$F_{el} = \left| e\vec{E}_e \right| = \frac{1}{4\pi\varepsilon_0} \frac{e^2}{d^2} \quad \text{to right}$$

Net Force:

$$\vec{F}_{net} = \vec{F}_{el} + \vec{F}_{mag}$$

$$F_x = +\frac{1}{4\pi\varepsilon_0} \frac{e^2}{d^2} - \left( \frac{\mu_0}{4\pi} \frac{e^2 v_p v_e}{d^2} \right) \sin\theta$$

$$= \frac{1}{4\pi\varepsilon_0} \frac{e^2}{d^2} \left[ 1 - \frac{v_e v_p \sin\vartheta}{c^2} \right] \quad (\text{since } \mu_0\varepsilon_0 = \frac{1}{c^2})$$

$$F_y = \left( \frac{\mu_0}{4\pi} \frac{e^2 v_p v_e}{d^2} \right) \cos\theta$$

(c) It is clear from the diagrams above that the net force on the proton is NOT equal and opposite to the net force on the electron! This is confirmed by the calculation; the x-components of the force are equal and opposite, but the y-components are not. Evidently Newton's third law is not obeyed by magnetic forces.

**HW12-2:**

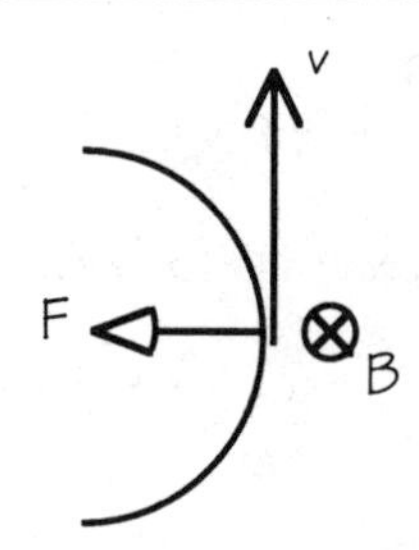


In region of circular motion, the magnetic force is toward center of circle:

$$\vec{F}_{mag} = q\vec{v}\times\vec{B}; \quad q = +e \text{ (positive)}$$

$\vec{F}$ is toward center of circle, so $\vec{B}$ is into page.

$$|m\vec{a}| = m\frac{v^2}{r} = |e\vec{v}\times\vec{B}| = evB$$

$$v = \frac{eBr}{m}$$

In deflection region, speed is same as in circular region;

net force is zero, so $F_{mag} = F_{el}$

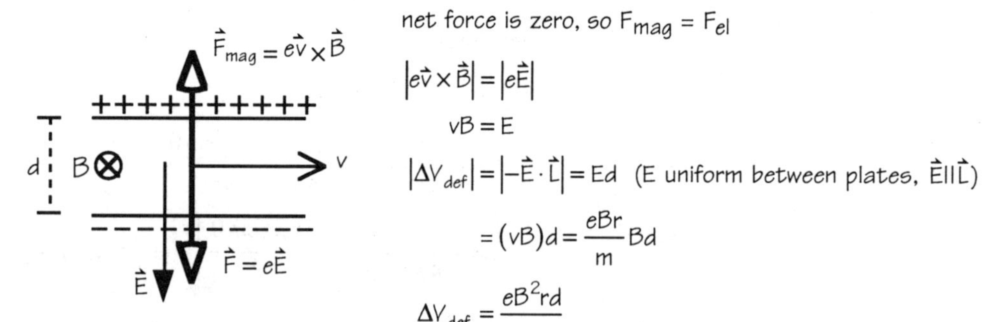


$$|e\vec{v}\times\vec{B}| = |e\vec{E}|$$

$$vB = E$$

$$|\Delta V_{def}| = |-\vec{E}\cdot\vec{L}| = Ed \quad (\text{E uniform between plates, } \vec{E}\parallel\vec{L})$$

$$= (vB)d = \frac{eBr}{m}Bd$$

$$\Delta V_{def} = \frac{eB^2rd}{m}$$

Electric field points down.

In acceleration region, ion accelerates from rest to final speed v

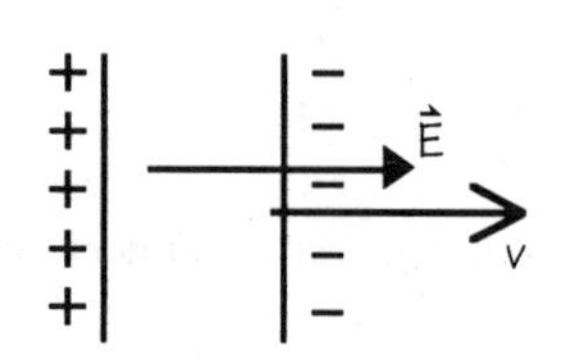


$$|\Delta PE| = |\Delta KE|$$

$$e\Delta V_{acc} = \left(\frac{1}{2}mv^2 - 0\right)$$

$$\Delta V_{acc} = \frac{mv^2}{2e} = \frac{eB^2r^2}{2m}$$

Electric field points in direction of v (to right).

(a)

$$^{12}C^+: \quad \Delta V_{acc} = \left(\frac{e}{m}\right)\frac{B^2r^2}{2} = \frac{(1.6\text{x}10^{-19}\ C)}{\left(\dfrac{12\text{x}10^{-3}\ \text{kg}}{6\text{x}10^{23}\ \text{atoms}}\right)}\frac{(0.2\ T)^2(0.1\ m)^2}{2} = 1600\ \text{volts}$$

$$\Delta V_{def} = \left(\frac{e}{m}\right)B^2rd = \frac{(1.6\text{x}10^{-19}\ C)}{\left(\dfrac{12\text{x}10^{-3}\ \text{kg}}{6\text{x}10^{23}\ \text{atoms}}\right)}(0.2\ T)^2(0.1\ m)(.01\ m) = 320\ \text{volts}$$

(b)

$$^{14}C^{+}:\ \Delta V_{acc} = \left(\frac{e}{m}\right)\frac{B^2r^2}{2} = \frac{\left(1.6\text{x}10^{-19}\ C\right)}{\left(\dfrac{14\text{x}10^{-3}\ kg}{6\text{x}10^{23}\ atoms}\right)}\frac{(0.2\ T)^2(0.1\ m)^2}{2} = 1371\ volts$$

$$\Delta V_{def} = \left(\frac{e}{m}\right)B^2rd = \frac{\left(1.6\text{x}10^{-19}\ C\right)}{\left(\dfrac{14\text{x}10^{-3}\ kg}{6\text{x}10^{23}\ atoms}\right)}(0.2\ T)^2(0.1\ m)(.01\ m) = 274\ volts$$

(c) Directions of fields are shown and explained above.

**HW12-3:**

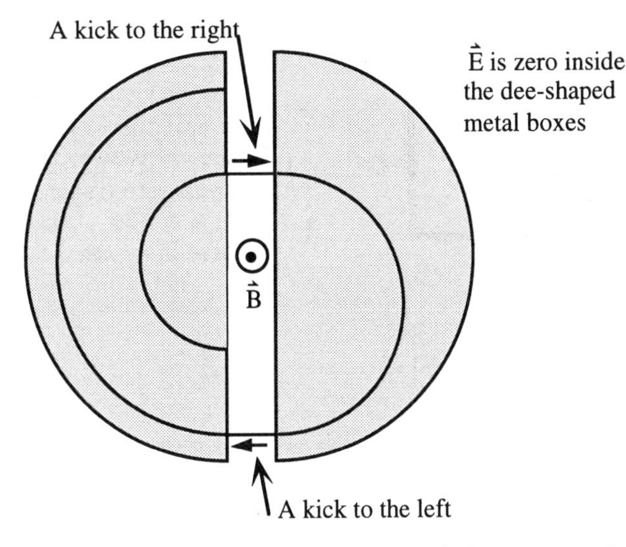


(a) The proton follows a circular orbit under the influence of the magnetic force, so

$$F_{magnetic} = evB\sin(90°) = evB = ma = m\frac{v^2}{r}$$

Solving for v, we have $v = \frac{eBr}{m} = \frac{2\pi r}{T}$, where T is the period—the time it takes to go around once.

Solve for the period: $$T = \frac{2\pi m}{eB} = \frac{2\pi\left(\dfrac{10^{-3}\text{kg / mole of atomic hydrogen}}{6\times10^{23}\text{protons / mole}}\right)}{\left(1.6\times10^{-19}C\right)(1\text{tesla})} = 6.5\times10^{-8}\ sec,$$

independent of r and v.

(b) The frequency $f = \frac{1}{T} = \frac{eB}{2\pi m} = \frac{1}{6.5\times10^{-8}\ sec} = 1.53\times10^7\text{hertz}.$

(c) Maximum speed is achieved at the outer radius R = 0.15 meters.

$$\Delta(KE) = \Delta\left(\frac{1}{2}mv^2\right) = e\Delta V_{equivalent}$$

$$\Delta V_{equivalent} = \frac{1}{2}\frac{m}{e}v^2 = \frac{1}{2}\frac{m}{e}\left(\frac{eBR}{m}\right)^2 = \frac{1}{2}\frac{e}{m}B^2R^2$$

$$\Delta V_{equivalent} = \frac{1}{2}\frac{\left(1.6\times10^{-19}C\right)}{\left(\frac{10^{-3}kg}{6\times10^{23}}\right)}(1tesla)^2(0.15m)^2 = 10^6\text{ volts}$$

(d) In each complete orbit there are two 500-volt potential drops, or 1000 volts per orbit, so it takes 1000 orbits to get to $10^6$ volts. Each orbit takes $6.5\times10^{-8}$ sec, so the total time to get from the center of the cyclotron to the outer edge is $1000\times6.5\times10^{-8}$ sec = $6.5\times10^{-5}$ sec, which is 65 microseconds.

**HW12-4:**

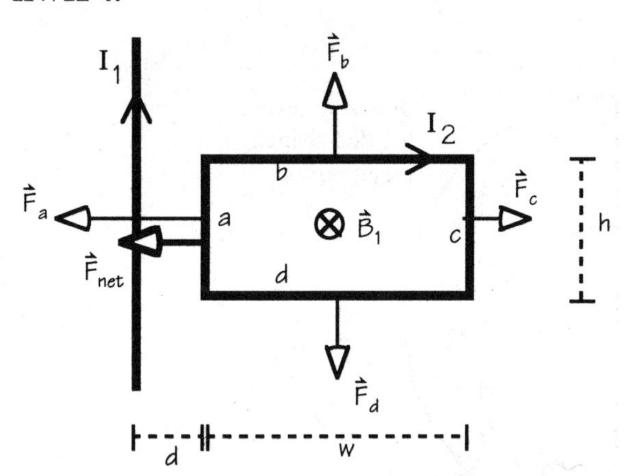


The magnetic field $B_1$ due to current $I_1$ is into the page in the region of the loop. However, the magnitude of $B_1$ varies with distance from the wire, so $B_1$ is smaller in magnitude along wire c than along wire a.

$\vec{F}_d + \vec{F}_b = 0$ (because, by symmetry, the forces on segments b & d are equal and opposite)

$$F_{net} = F_a - F_c = \left| I_2 \vec{L}_a \times \vec{B}_{1\ \text{at a}} - I_2 \vec{L}_c \times \vec{B}_{1\ \text{at c}} \right|$$

$$= I_2 h \frac{\mu_0}{4\pi}\frac{2I_1}{d} - I_2 h \frac{\mu_0}{4\pi}\frac{2I_1}{d+w}$$

$$F_{net} = \frac{\mu_0}{4\pi}\left(\frac{2I_2I_1hw}{d(d+w)}\right) \text{ to the left}$$

**HW12-6:**

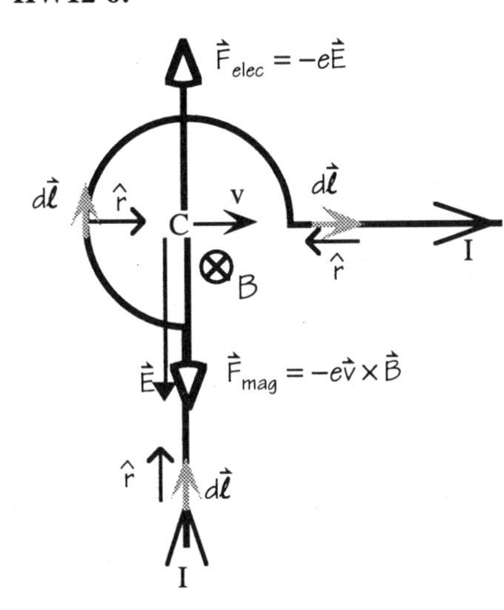


Directions:

The magnetic field created by the current is into the page. The loop contributes B into the page, and the straight segments make no contribution, since for each one $d\vec{\ell} \times \hat{r} = 0$. (see diagram).

The magnetic force on the electron is therefore down, as shown (because the particle is negative).

Net force = 0 so electric force is up, as shown.

Electric field points down, as indicated in diagram..

Magnitude of E:

$$F_{mag} = F_{el}$$

$$\left|-e\vec{v} \times \vec{B}\right| = \left|-e\vec{E}\right|$$

$$vB = E$$

$$B = \left| \int_{\frac{3}{4}\text{loop}} \frac{\mu_0}{4\pi} \frac{I\,d\vec{\ell} \times \hat{r}}{R^2} \right| = \frac{\mu_0}{4\pi} \frac{I}{R^2} \int_{\frac{3}{4}\text{loop}} d\ell = \frac{\mu_0}{4\pi} \frac{I}{R^2} \left( \frac{3}{4} 2\pi R \right) = \frac{3}{8} \frac{\mu_0 I}{R}$$

$$E = \frac{3}{8} \frac{\mu_0 I v}{R}$$

**HW12-8:**

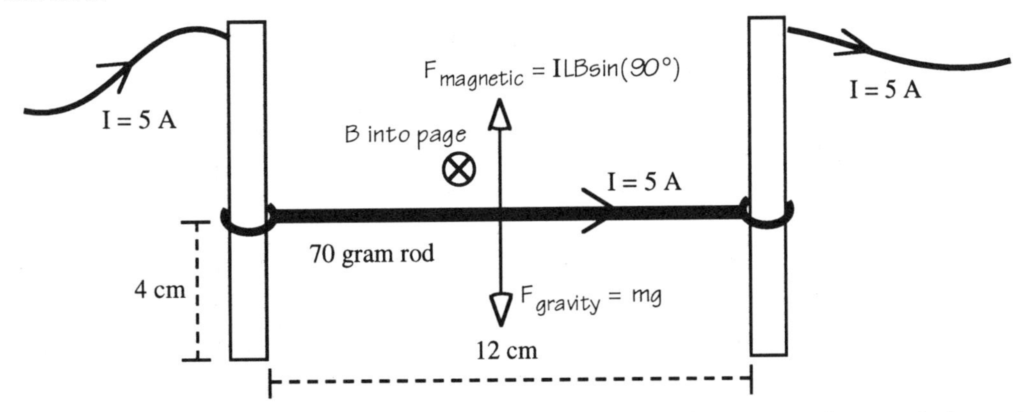


In order to balance the downward force of gravity, there must be an upward magnetic force of equal magnitude, which requires a magnetic field into the page (right hand rule, with conventional current to the right).

$$ILB\sin(90°) = ILB = mg$$

$$B = \frac{mg}{IL} = \frac{(0.07\text{kg})(9.8\text{N/kg})}{(5\text{ampere})(0.12\text{m})} = 1.14\text{ tesla}$$

There could be an additional component of magnetic field to the right or left, which would exert no force on the wire: $\left|I\vec{L} \times \vec{B}\right| = ILB\sin(0°) = 0$.

**HW12-10:**

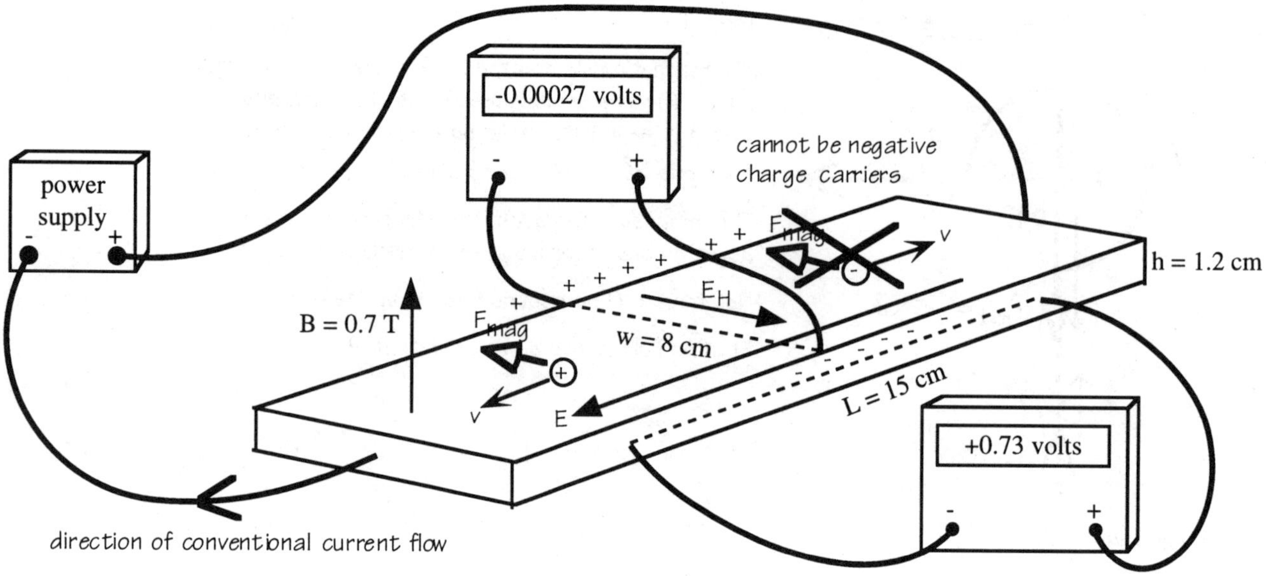

(a) The voltmeter reading of -0.00027 volts indicates that the back side of the slab is at a higher potential than the front side, so there is a Hall-effect transverse electric field $E_H$ as shown. That means that there must be extra + charge on the back side (and extra - charge on the front).

The conventional current flow and the main electric field E point as shown, and the +0.73 volt reading is consistent: the potential is dropping in the direction of E and the conventional current.

Positive carriers move in the direction of conventional current, and experience a magnetic force toward the back side, which would lead to + charge buildup on the back side, which is what is observed.

Negative carriers move opposite to the conventional current, and would also experience a magnetic force toward the back side, which would lead to - charge buildup on the back side, which is not observed.

So the charge carriers are positive.

(b) In the steady state, the transverse electric and magnetic forces must balance, so $qE_H = qvB\sin(90°) = qvB$. So $E_H = vB$. The drift speed v is uniform throughout the slab (current conservation and constant cross-sectional area), and the magnetic field B is uniform throughout this region. So $E_H = vB$ must be uniform along the 8-cm path across the slab, and we can write

$$|\Delta V| = 0.00027 \text{ volts} = \left| -\int_i^f \vec{E} \bullet d\vec{l} \right| = E_H w = E_H(0.08\text{m})$$

$$E_H = \frac{0.00027 \text{ volts}}{0.08\text{m}} = 0.0034 \text{ volts / meter}$$

$$v = \frac{E_H}{B} = \frac{\left(\frac{0.00027 \text{ volts}}{0.08\text{m}}\right)}{0.7\text{tesla}} = 4.8 \times 10^{-3} \text{m / s}$$

Note that we have experimentally determined the drift speed v, independent of the carrier charge q and the density of charge carriers n.

(c) $v = uE$, where E is the electric field in the direction of the current. Since v is uniform, E must be uniform, and we can write

$$|\Delta V| = 0.73\,\text{volts} = \left|-\int_i^f \vec{E} \bullet d\vec{l}\right| = EL = E(0.15\text{m})$$

$$E = \frac{0.73\text{volts}}{0.15\text{m}} = 4.9\,\text{volts / meter}$$

$$u = \frac{v}{E} = \frac{4.8 \times 10^{-3}\,\text{m / s}}{\left(\dfrac{0.73\text{volts}}{0.15\text{m}}\right)} = 9.9 \times 10^{-4}\,\frac{\text{m / s}}{\text{volts / m}}$$

Note that the electric field E in the direction of the current is much larger than the transverse field $E_H$.

(d) $I = qnAv$, and we assume that $q = e$. So we have

$$n = \frac{I}{eAv} = \frac{0.3\text{ampere}}{(1.6 \times 10^{-19}\text{C})(0.08\text{m} \times 0.12\text{m})(4.8 \times 10^{-3}\,\text{m / s})} = 4 \times 10^{23}\,\text{carriers / m}^3$$

This is a very low density of charge carriers. The density of free electrons in copper is $8 \times 10^{28}$ per $m^3$ (see page 208). Evidently this slab of material is not an ordinary metal: the charge carriers are positive, and the density of charge carriers is very low.

(e) The potential difference along 15 cm of material is 0.73 volts = RI, so

$$R = \frac{\Delta V}{I} = \frac{0.73\text{volts}}{0.3\text{ampere}} = 2.4\,\text{ohms}$$

## Chapter 13 solutions

**RQ13-4: (with grading key: 10 pts)**

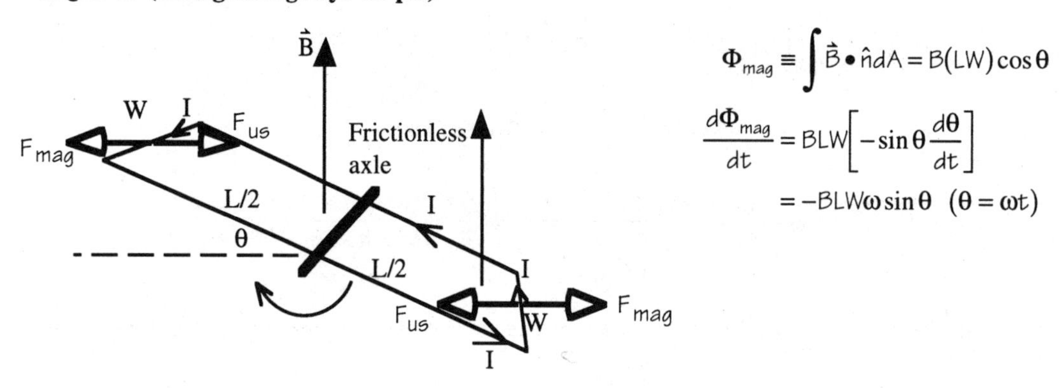


Magnitude of the emf when loop is at an angle $\theta$ to the horizontal is $|emf| = BLW\omega|\sin\theta|$ (4 pts).

At the instant shown in the diagram, the flux through the loop is decreasing, so the current I is counter-clockwise to attempt to make up for the decreasing flux (3 pts).

We have to apply forces ($F_{us}$ on the diagram) to oppose the $I\vec{L} \times \vec{B}$ magnetic forces on the current-carrying wires. We twist the loop clockwise against the counter-clockwise twist due to the magnetic forces (3 pts).

**RQ13-5: (with grading key: 10 pts)**

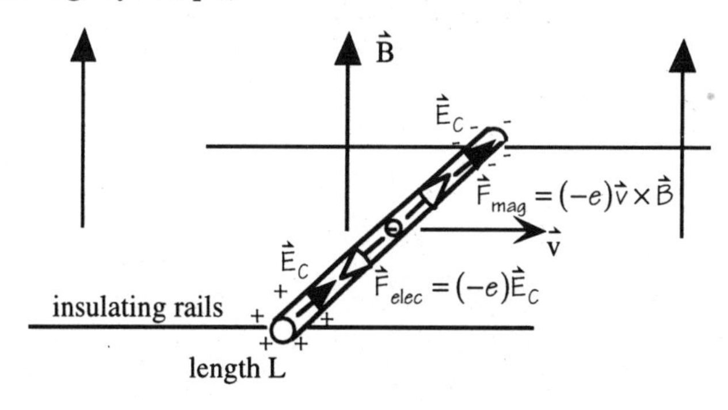


(1 pt for surface charge)

Magnetic force $(-e)\vec{v} \times \vec{B}$ on an electron inside the rod polarizes the rod as shown (1 pt).

A Coulomb electric field $\vec{E}_C$ appears inside the rod as shown, due to the surface charges (1 pt).

Polarization proceeds until $F_{electric} = F_{magnetic}$:

$$eE_C = evB\sin(90°) \Rightarrow E_C = vB \quad \text{(2 pts)}$$

$\vec{E}_C$ is uniform throughout the rod, since $\vec{v}$ and $\vec{B}$ are the same everywhere. The surface charges will arrange themselves to make a uniform electric field.

$\Delta V \equiv -\int \vec{E}_C \bullet d\vec{l} = E_C L = vBL$, since $\vec{E}_C$ is uniform (2 pts).

$\Delta V = emf - r_{internal} I = emf$, since I = 0, so emf = $E_C L$. = vBL (1 pt).

Or, $emf = \int \frac{\vec{F}_{NC} \bullet d\vec{l}}{q} = \frac{evBL}{e} = vBL = E_C L$.

Once the rod is polarized, I = 0, so there is no ILB magnetic force opposing the motion. So no force is required to keep the rod moving at a constant speed v on the frictionless rails (2 pts).

**HW13-3:**

(a)

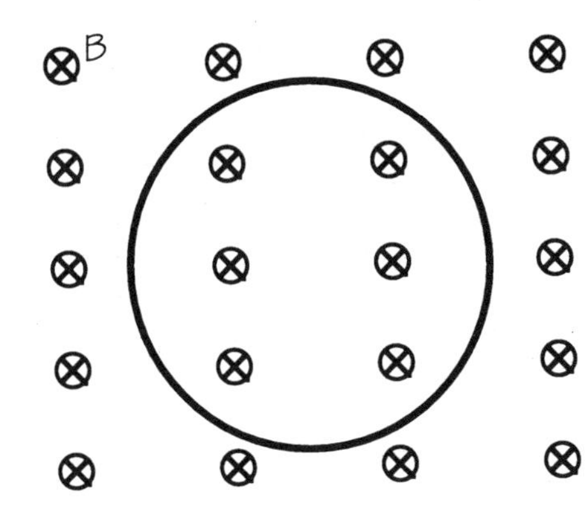


Consider a loop of wire lying in a region where the magnetic field is uniform and into the page.

B decreases with time ⇒ magnetic flux decreases

Decrease in flux ⇒ current runs (clockwise) to make B into page to attempt to keep flux constant.

So induced field is in same direction as applied field.

(b)

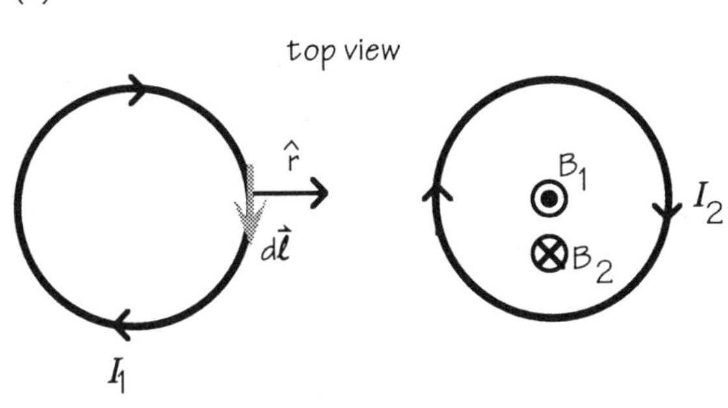


Inside ring 2, $B_1$ points out of the page because the nearer side of ring 1 (producing B out) contributes more than the farther side (which produces B in).

As $I_1$ increases, $B_1$ increases, so flux "out of the page" inside ring 2 increases.

The induced current in ring 2 must make a magnetic field $B_2$ into the page inside ring 2, to try to oppose the increase in flux, so $I_2$ runs clockwise as well.

(c)

end-on view

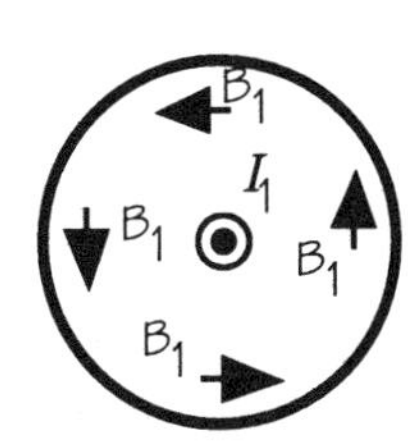


Flux through ring $= \int_{\substack{\text{bounded}\\\text{surface}}} \vec{B}_1 \cdot \hat{n}\, dA = 0$ because $\vec{B}_1 \perp \hat{n}$ so $\vec{B}_1 \cdot \hat{n} = 0$.

So $\dfrac{d\Phi}{dt} = 0$ and emf $= 0$ and $I = 0$.

(d) Changing magnetic field induces currents to run around the copper tube:

B due to magnet

S

N

Here B is decreasing, induces current to run clockwise to oppose flux change. This induced current makes a magnetic field pointing down at the location of the magnet. So induced current loop acts like a magnet with N on the bottom, which attracts the falling magnet, slowing it down.

Here B increasing, induces current to run counter-clockwise to oppose flux change. This induced current makes a magnetic field pointing up at the location of the magnet,. So induced current loop acts like a magnet with N on the top, which repels the falling magnet, slowing it down.

B due to magnet

(e) If the solenoid is filled with iron, the magnetic field inside would be much larger for a given current. Thus, changes in the magnetic field (and hence the magnetic flux) would be larger for a given change in current, so its inductance would be increased.

**HW13-4:**

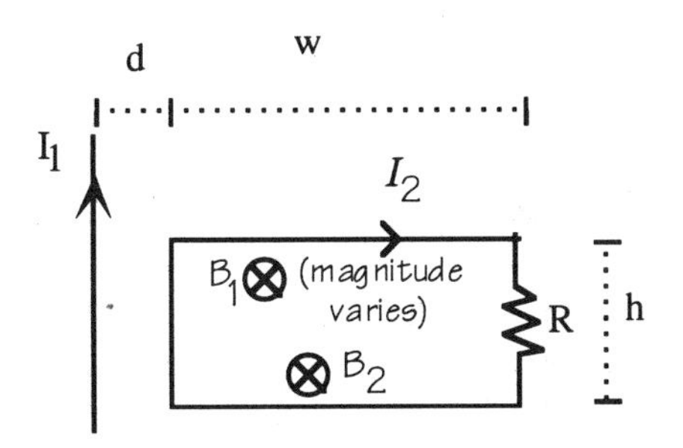


(a) The current in the long wire makes a magnetic field into the page everywhere inside the loop. As the current decreases, the flux through the loop decreases, and a current $I_2$ runs in the loop to produce $B_2$ into the page, to oppose the change in flux. The current in the loop therefore runs clockwise.

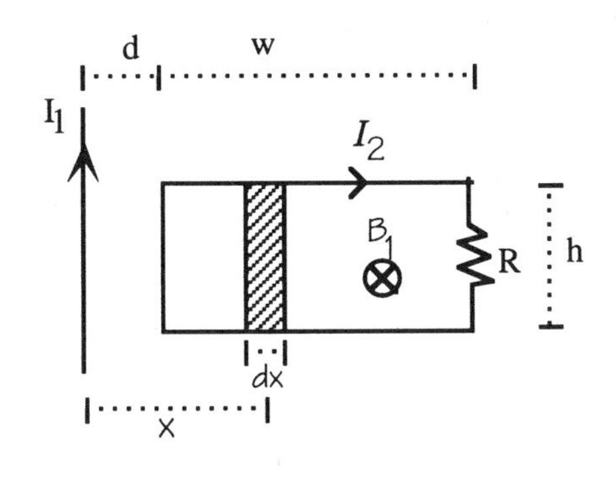


(b) Flux through a narrow strip of area (as shown) is:

$$d\Phi = (B_{1 \text{ at } x})(dA) = \left(\frac{\mu_0}{4\pi}\frac{2I_1}{x}\right)(h)(dx)$$

Net flux through loop at a given time is:

$$\Phi = \int_d^{d+w} d\Phi = \int_d^{d+w} \left(\frac{\mu_0}{4\pi}\frac{2I_1 h}{x}\right)dx$$

$$= \frac{\mu_0}{4\pi}2I_1 h \int_d^{d+w}\frac{dx}{x} = \frac{\mu_0}{4\pi}2I_1 h[\ln x]_d^{d+w}$$

$$= \frac{\mu_0}{4\pi}2I_1 h\left(\ln\frac{d+w}{d}\right)$$

Time rate of change of flux gives emf:

$$|emf_2| = \left|\frac{d\Phi}{dt}\right| = \left|\frac{d}{dt}\left(\frac{\mu_0}{4\pi}2I_1 h\left(\ln\frac{d+w}{d}\right)\right)\right|$$

$$= \left(\frac{\mu_0}{4\pi}2h\left(\ln\frac{d+w}{d}\right)\right)\left|\frac{dI_1}{dt}\right|$$

Current:

$$I_2 = \frac{emf_2}{R} = \frac{\frac{\mu_0}{4\pi}2h\left(\ln\frac{d+w}{d}\right)\left|\frac{dI_1}{dt}\right|}{R}$$

**HW13-5:**

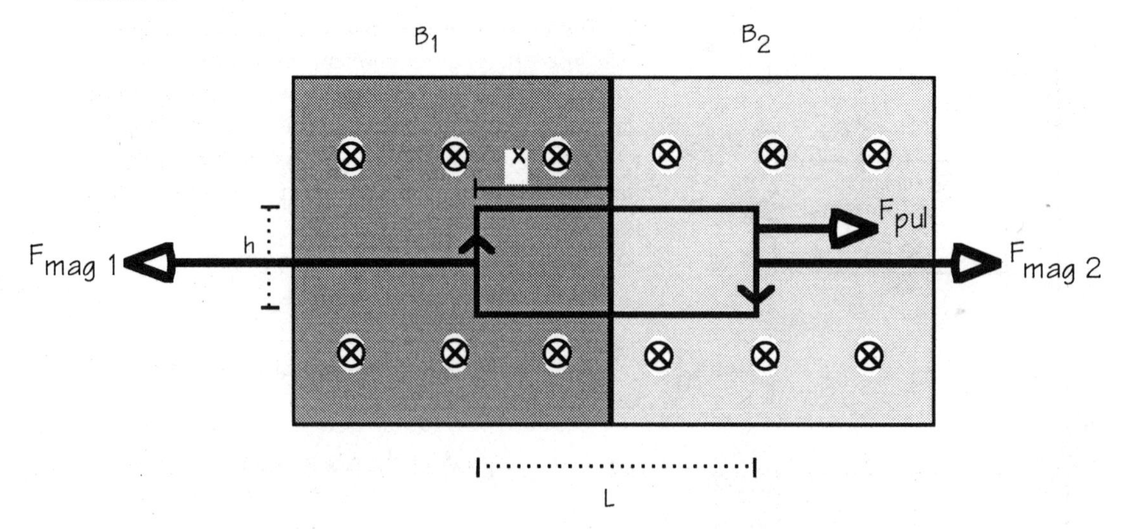


Since $B_2 < B_1$, the flux through the loop is decreasing as the loop moves to the right. To oppose this decrease, a clockwise current is induced in the loop, making $B_{loop}$ into the page. The current-carrying wires experience magnetic forces. The magnetic forces on the horizontal segments are equal and opposite, but the magnetic force (to the right) on the right end is smaller than the magnetic force (to the left) on the left end. An external pulling force must be exerted to balance the total magnetic force so the net force will be zero and the loop will travel at constant speed.

**Induced emf:**

$$\Phi = \sum \vec{B}\cdot\hat{n}\Delta A = (B_1hx) + (B_2h(L-x)) = (B_1h - B_2h)x + B_2hL$$

$$|emf| = \left|\frac{d\Phi}{dt}\right| = \left|\frac{d}{dt}\left[(B_1h-B_2h)x + B_2hL\right]\right| = \left|(B_1h-B_2h)\frac{dx}{dt}\right| = \left|(B_1h-B_2h)v\right|$$

**Current:**

$$I = \frac{|emf|}{R} = \frac{(B_1h-B_2h)v}{R}$$

**Magnetic Force:**

$$\vec{F}_{mag} = I\vec{L}_1\times\vec{B}_1 + I\vec{L}_2\times\vec{B}_2$$

$$F_{mag} = I(B_1h - B_2h) \text{ to the left}$$

**Pulling Force:**

$$F_{pull} = F_{mag} = \frac{(B_1h-B_2h)v}{R}(B_1h-B_2h)$$

$$F_{pull} = \frac{(B_1-B_2)^2h^2v}{R} \text{ to the right}$$

**Speed:**

$$v = \frac{F_{pull}R}{(B_1-B_2)^2h^2}$$

**HW13-6:**

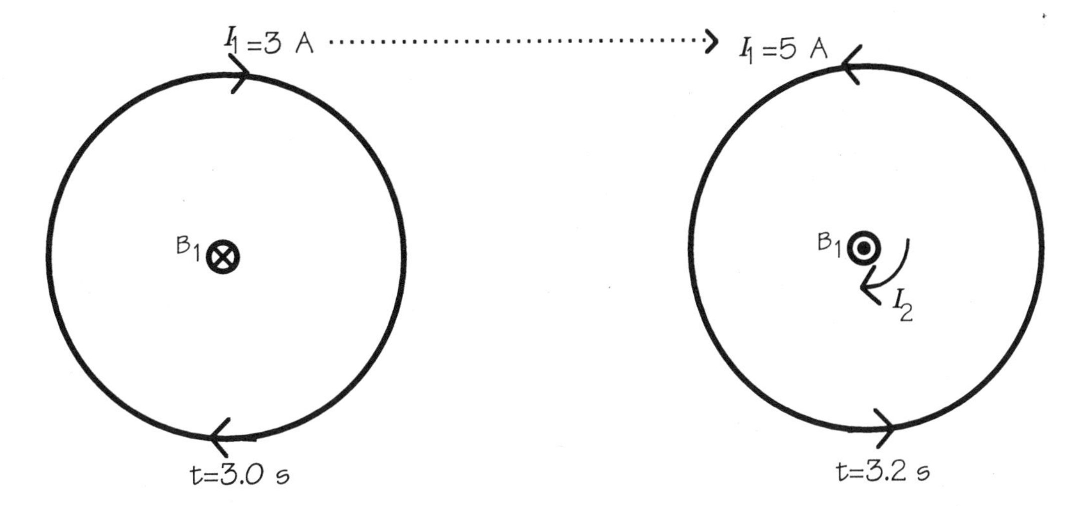


Initially $B_1$ (magnetic field due to large ring, at location of small ring) is into the page; after 0.2 seconds it is out of the page, so $B_1$ is increasing out of the page. This induces a current in the small ring in a direction to make $B_2$ into the page, to oppose the increase in flux, so a clockwise current $I_2$ runs in the small ring during the time that $I_1$ is changing.

Since B1 is nearly uniform over the small area of the small ring,

$$\Phi \approx B_1 A_{\substack{\text{small}\\\text{ring}}} = \left(\frac{\mu_0}{4\pi}\frac{2\pi I_1}{R}\right)(\pi r^2)$$

(of course this value changes with time, since $I_1$ changes with time.)

$$|\text{emf}| = \left|\frac{\Delta\Phi}{\Delta t}\right| = \left|\frac{\left(\left(\frac{\mu_0}{4\pi}\frac{2\pi I_{1\,3.0\text{ s}}}{R}\right) - \left(\frac{\mu_0}{4\pi}\frac{2\pi I_{1\,3.2\text{ s}}}{R}\right)\right)\left(A_{\substack{\text{small}\\\text{ring}}}\right)}{\Delta t}\right|$$

$$= \left|\frac{\frac{\mu_0}{2(.5\text{ m})}(+5\text{ A}-(-3\text{ A}))\left(\pi(.005\text{ m})^2\right)}{(3.2\text{ s}-3.0\text{ s})}\right|$$

$$|\text{emf}| = 3.9\text{x}10^{-9}\text{ volts}$$

**HW13-7:**

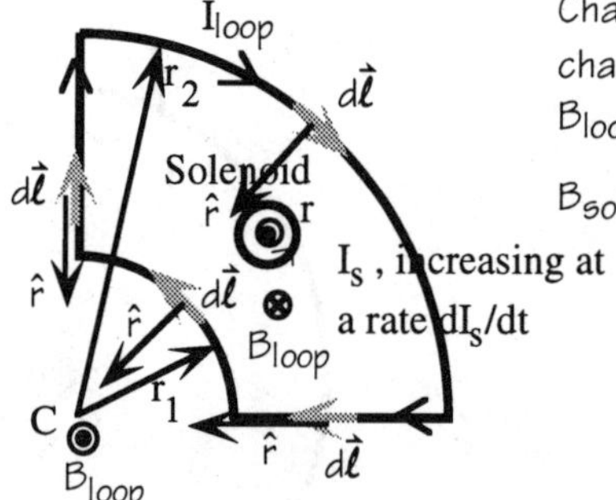


Changing $I_{solenoid} \Rightarrow$ changing $B_{solenoid} \Rightarrow$ changing $\Phi \Rightarrow$ emf in outer loop $\Rightarrow I_{loop} \Rightarrow B_{loop}$ at location C.

$B_{solenoid} \approx 0$ outside solenoid.

1) Inside solenoid, $B_{solenoid} = \mu_O \frac{N}{L} I_{solenoid}$ out of page.

2) Flux through outer loop = $B_{solenoid} A_{solenoid} \cos(0°) + (\approx 0) A_{rest\ of\ loop} = \mu_O \frac{N}{L} I_{solenoid} \left(\pi r^2\right)$

3) Flux increasing out of page; induced current $I_{loop}$ makes $B_{loop}$ into page, so current clockwise as shown.

$$|emf| = \left|\frac{d\Phi}{dt}\right| = \frac{d}{dt}\left(\mu_O \frac{N}{L} I_{solenoid}\right)\left(\pi r^2\right) = \mu_O \frac{N}{L} \pi r^2 \frac{dI_{solenoid}}{dt}$$

$$I_{loop} = \frac{emf}{R} = \mu_O \frac{N}{L} \frac{\pi r^2}{R} \frac{dI_{solenoid}}{dt}$$

4) $B_{at\ C} = B_{straight\ segments} + B_{inner\ wire} + B_{outer\ wire}$

$B_{straight\ segments} = 0$ because $d\vec{\ell} \times \hat{r} = 0$ (see diagram)

$$\vec{B}_{inner\ wire} = \left| \int_{quarter\ loop} \frac{\mu_O}{4\pi} \frac{I_{loop} d\vec{\ell} \times \hat{r}}{r_1^2} \right| = \frac{\mu_O}{4\pi} \frac{I_{loop}}{r_1^2} \int_{quarter\ loop} d\ell \sin(90°)$$

$$= \frac{\mu_O}{4\pi} \frac{I_{loop}}{r_1^2} \left(\frac{2\pi r_1}{4}\right) = \frac{\mu_O}{4\pi} \frac{I_{loop}}{2r_1}$$

$B_{inner\ wire}$ is out of the page, and $B_{outer\ wire}$ is into the page at location C. Since B is proportional to 1/r, we see that $B_{inner\ wire}$ is larger than $B_{outer\ wire}$. So the net field is out of the page, with magnitude

$$B_{net\ at\ C} = B_{inner\ wire} - B_{outer\ wire}$$

$$= \frac{\mu_O}{4\pi} \frac{I_{loop}}{2} \left(\frac{1}{r_1} - \frac{1}{r_2}\right)$$

$$= \frac{\mu_O}{4\pi}\left[\mu_O \frac{N}{L} \frac{\pi r^2}{2R} \frac{dI_{solenoid}}{dt}\right]\left(\frac{1}{r_1} - \frac{1}{r_2}\right) \text{ out of page}$$